Radiation Technology
for Polymers

Radiation Technology for Polymers

Third Edition

Jiri George Drobny

CRC Press
Taylor & Francis Group
Boca Raton London New York

CRC Press is an imprint of the
Taylor & Francis Group, an **informa** business

Third edition published 2020
by CRC Press
6000 Broken Sound Parkway NW, Suite 300, Boca Raton, FL 33487-2742

and by CRC Press
2 Park Square, Milton Park, Abingdon, Oxon, OX14 4RN

© 2021 Taylor & Francis Group, LLC

CRC Press is an imprint of Taylor & Francis Group, LLC

First published by CRC Press in 2002
Second edition 2010

ISBN: 978-0-367-18932-7 (hbk)
ISBN: 978-0-367-51192-0 (pbk)
ISBN: 978-0-429-20119-6 (ebk)

Typeset in Palatino
by Lumina Datamatics Limited

To Betty Anne

Contents

Preface to First Edition

The industrial use of ultraviolet (UV) and electron beam (EB) radiation is growing at a fast pace and is penetrating many areas, such as electronics, automotive, printing, adhesives and coatings, packaging, etc., which traditionally have had their own well-established processes. Information on this topic useful to a professional can be found in many places, such as encyclopedias (Encyclopedia of Polymer Science and Engineering, Ullmann's Encyclopedia of Industrial Chemistry), professional publications (PCI, Paint and Coating Industry, Paint & Powder, FLEXO, Converting, Modern Plastics, Rubber Chemistry and Technology, Modern Plastics, Wire and Cable, Radiation Physics and Chemistry, Journal of Applied Polymer Science) and others. RadTech News, a publication of RadTech North America covers applications, new technology, and industry news. During the past few years, several very informative books have been published by SITA Technology Ltd. in the UK that covers different aspects of UV/EB radiation technology in great detail. However, seeking specific information may be prohibitively time consuming and a need for a quick reference book is obvious.

Radiation Technology of Polymers is designed to meet this need by providing systematic fundamental information about the practical aspects of UV/EB radiation to professionals in many different fields. The intended audiences are mainly chemists or chemical engineers new to UV/EB radiation technology. Another reader of this book may be a product or process designer looking for specifics about the effects of UV/EB on a specific polymeric material or for a potential technological tool. This book may also be a useful resource for recent college and university graduates or for graduate or undergraduate students pursuing polymer science and engineering and people who undergo corporate training. Given the breadth of the field and the multitude of applications available, the book does not go into details; this is left to publications of a much larger size and scope and to professional periodicals. Rather, it covers the essentials and points the reader toward sources of more specific and/or detailed information. *Radiation Technology of Polymers* is not a competition to other books on the subject, it merely complements them. With this in mind, this book is divided into 11 separate chapters, covering the principles of generating UV and EB energy, equipment, processes, applications, dosimetry, safety and hygiene. The last chapter covers the newest developments and trends.

This book began as lectures and seminars at the Plastics Engineering Department of the University of Massachusetts at Lowell and to varied professional groups and companies in the United States and abroad. It draws from the author's more than 40 years of experience as an R&D and manufacturing professional in the polymer processing industry and more recently as an independent international consultant.

My thanks are due to the team from the CRC Press, Susan Farmer, Helena Redshaw and Sylvia Wood for helping to bring this work to fruition, to my family for continuing support and Dr. Ewa Andrzejewska, John Chrusciel, Dr. Joseph Koleske, and Richard W. Stowe for their valuable comments and recommendations.

Merrimack, New Hampshire

Preface to Second Edition

The first edition of *Radiation Technology for Polymers* had the main goal to provide systematic fundamental information to professionals entering the industrial practice of radiation technology as it applies to processing of polymers or people already working in this field. Since its publication in 2003, the industry has changed markedly. Many technological developments have taken place, new applications and products have been developed and commercialized, and some already established ones were discontinued. Companies were sold and bought, reorganized and/or renamed. The UV/EB technology is becoming increasingly more important in a variety of issues, such as continuing quest for further reduction of volatile organic compounds and toxic substances in the environment, development of alternative sources of energy and more. Thus, it was time to update the publication to include these changes, developments, and issues. As the first edition, the second edition still emphasizes the practical aspects of UV/EB technology and its industrial applications. Few illustrations were added and one of the major features is the addition of processing and engineering data of some commercial products. In addition, the feedback from colleagues, students, clients and attendants of various seminars and training sessions were helpful in preparing the manuscript of this updated and expanded edition.

Special thanks go to Anthony J. Berejka, Dr. Marshall R. Cleland, and Richard W. Stowe for editing parts of the manuscript and providing helpful comments and recommendations.

Allison Shatkin who was very helpful and encouraging from the beginning until the end of the preparation of the manuscript deserves many thanks. A special credit is due to the team from CRC Press, particularly to Andrea Dale, Jennifer Ahringer, and Amy Rodriguez for bringing this work to fruition.

Jiri George Drobny
Merrimack, NH and Prague, Czech Republic

Preface to Third Edition

The previous two editions of *Radiation Technology for Polymers* had as the main goal to provide systematic, fundamental information to professionals who are entering the industrial practice of radiation technology as it applies to processing of polymers, and to those already working in this field. Since the publication of the second edition in 2010, many technological developments and changes have taken place; new applications and products have been developed and commercialized, and some already established ones were discontinued; companies have been sold and bought, reorganized, and/or renamed.

UV/EB technology continues to be important in a variety of issues, such as the continuing quest for further reduction of volatile organic compounds and toxic substances in the environment, development of alternative sources of energy, and more. Thus, it was time to update the publication to include these changes, developments, and issues.

As the previous editions, the third edition still stresses the practical aspects of UV/EB technology and its industrial applications. In order to cover the technological progress and other changes better, a contributing author, a specialist in specific areas, is covering individual chapters or chapter sections. The original author has taken the responsibility as editor of the entire work in addition to writing several chapters that cover areas of his expertise. As before, a few illustrations were added, and the processing and engineering data of selected commercial products were updated. One of the major challenges was the fact that most of the conventional equipment and technologies are still used, so we still had to keep the appropriate information available and yet insert the new items so that it does not become incoherent or confusing.

We have continued to obtain feedback from colleagues, students, clients, and attendants of various seminars and training sessions as an important source for preparing the manuscript of this updated and expanded edition. There were, as always, good friends and colleagues who helped by giving valuable advice, providing photographs, data, and connections within the industry. We are very thankful to John Chrusciel, David Engberg, Michael Fletcher, Mickey Fortune, Georgina Gonzales, Stacey Hoge, Kevin Joesel, Urs Läuppi, Bengt Laurell, Erich Midlik, and Karl Swanson for doing just that. The very rapid and dynamic growth of the industry presented a major challenge, and we hope that we are providing a true and accurate snapshot of the current state of the knowledge and technology.

As in the case of the two previous editions, Allison Shatkin from CRC Press was very helpful and encouraging from the beginning until the end of the preparation of the manuscript and deserves many thanks. A special credit is due to the team from CRC Press, particularly to Gabrielle Vernachio for bringing this work to fruition.

Jiri George Drobny
Merrimack, NH and Prague, Czech Republic

Author

Jiri George Drobny was educated at the Technical University in Prague in Chemical Engineering, at the Institute of Polymer Science of the University of Akron, and at Shippensburg State University in Business Administration. His career spans over 50 years in the rubber and plastics processing industry in Europe, the US, and Canada, mainly in R&D with senior and executive responsibilities. Currently, he is President of Drobny Polymer Associates, an international consulting firm specializing in fluoropolymer science and technology, radiation processing, and elastomer science and technology. He has been active as an educator, lecturer, author, and as a technical and scientific translator. He is a member of the Society of Plastics Engineers, American Chemical Society, and RadTech International North America and is listed in Who's Who in America, Who's Who in Plastics and Polymers, Who's Who in Science and Engineering, and Who's Who in the East. He resides in New Hampshire.

Contributing author: Ruben Rivera, Regional Sales Manager, South, Phoseon Technologies

1

Introduction

Radiant energy is one of the most abundant forms of energy available to humankind. Nature provides sunlight, the type of radiation that is essential for many forms of life and growth. There are some natural substances that generate yet another kind of radiation that can be destructive to life, but when harnessed it can provide other forms of energy or serve in medicine or industrial applications. These are radioactive substances that produce ionizing radiation by radioactive decay, which is the spontaneous breakdown of an atomic nucleus resulting in the release of energy and matter from the nucleus.

Human genius created its own devices for generating radiant energy useful in a great variety of scientific, industrial, and medical applications. Cathode ray tubes emit impulses that activate screens of computer monitors and televisions. X-rays are used not only as a diagnostic tool in medicine, but also as an analytical tool in inspection of manufactured products such as tires and other composite structures. Microwaves are used not only in cooking or as means of heating rubber or plastics, but also in a variety of electronic applications. Infrared radiation is used in heating, analytical chemistry, and electronics. Manmade ultraviolet radiation has been in use for decades in medical applications, analytical chemistry, and in a variety of industrial applications. Devices used to generate accelerated particles are not only valuable scientific tools but also important sources of ionizing radiation for industrial applications. Both ultraviolet (UV) and electron beam (EB) radiations are classified as electromagnetic radiations, along with infrared (IR) and microwave (MW). The differences between them in frequency and wavelength are in Table 1.1.

Polymeric substances, which are predominantly high molecular weight organic compounds, such as plastics and elastomers (rubber), respond to radiation in several ways. They may be gradually destroyed by UV radiation from sunshine when exposed for extended periods of time outdoors; they may more or less change their properties. On the other hand, manmade UV radiation is actually used to produce polymers from monomers (low-molecular-weight building blocks for polymers) or from oligomers (essentially very low-molecular-weight polymers). In these reactions, almost always a liquid is converted into a solid almost instantaneously. Ionizing radiation (γ-rays and high-energy electrons) is even more versatile: it is capable of converting monomeric and oligomeric liquids into solids but can also produce major changes in properties of solid polymers.

TABLE 1.1

Frequency and Wavelength of Various Types of Electromagnetic Radiation

Radiation	Wavelength, μm	Frequency, Hz
Infrared	$1-10^2$	$10^{15}-10^{12}$
Ultraviolet	$10^{-2}-1$	$10^{17}-10^{15}$
Microwave	10^3-10^5	$10^{12}-10^{10}$
X-rays soft	$10^{-2}-10^{-3}$	$10^{17}-10^{16}$
X-rays hard	$10^{-4}-10^{-3}$	$10^{19}-10^{17}$
Electron beam	$10^{-7}-10^{-5}$	$10^{21}-10^{18}$
Gamma rays	$10^{-6}-10^{-5}$	$10^{-20}-10^{18}$

1.1 Basic Concepts

Industrial applications involving large volume radiation processing of monomeric, oligomeric, and polymeric substances depend essentially on two electrically generated sources of radiation: accelerated electrons and photons from high-intensity ultraviolet lamps. The difference between these two is that accelerated electrons can penetrate matter and are stopped only by mass, whereas high-intensity UV light affects only surface.[1] Generally, processing of monomers, oligomers, and polymers by irradiation by UV light and electron beam is referred to as *curing*. This term encompasses chemical reactions including polymerization, cross-linking, surface modification, and grafting. These reactions will be discussed in detail in the appropriate chapters.

The process of conversion of liquid to solid is mainly designed to use on compositions based on nonvolatile monomers and oligomers with molecular weights of less than 10,000. These have low enough viscosities so that they can be applied without the use of volatile solvents (volatile organic compounds [VOCs]).[1]

This, of course, is greatly beneficial for the environment, more specifically, the air. In fact, some states have recognized UV/EB curing of coatings, printing inks, paints, and adhesives as environmentally friendly in their legislative actions.[1] For example, when the EB process is compared to solvent-based coatings and even to water-based technology, another "green" alternative to VOC-based technology, it is found to be far superior in energy consumption (see Table 1.2). Moreover, the UV/EB technology is superior to solvent-based systems concerning greenhouse gas emissions. Thus, the amount of CO_2 from solvent combustion, typically 37 g/m^3, generates a facility emission potential of 415.8 kg/h.[2]

Clearly, UV and EB radiation have a great deal in common, as shown above. However, there are also differences. Besides the nature of interacting

TABLE 1.2

Energy Needed to Dry/Cure Coatings

Property/Energy	System		
	Solvent	Water	EB-Curable
Solids, weight %	40	40	100
Diluent	Toluene	Water	None
Boiling point, °C	111	100	N.A.
Vapor Pressure, mm Hg (20°C)	22	17	N.A.
Energy to dry/cure, 1 g dried coating, J/g	555	3390	30

Source: Berejka, A.J., in *Proceedings 3rd Annual Green Chemistry and Engineering Conference: Moving Toward Industrial Ecology,* June 29–July 1, Washington, DC, p. 35, 1999; Berejka, A.J., *RadTech Report,* 17, September/October, p. 47, 2003.

with matter, where high-energy electrons penetrate, and photons cause only surface effects, there are issues concerning the capital investment and chemistry involved.

Without any doubt, the UV irradiation process is the lower-cost option, since the equipment is simpler, smaller, and considerably less expensive to purchase and operate. Normally, there is no need to use nitrogen inerting, which adds to operating expenses. However, free radical-based UV curing requires the addition of photoinitiators, some of which are expensive and may bring about some undesirable effects, such as discoloring of the film and often also odor. Both these effects can be minimized or eliminated by nitrogen blanketing.[3] It is much more difficult to cure pigmented films, particularly thick ones using UV irradiation, while these can be cured without problems by electron beams. Photoinitiator and sensitizer residues from UV-cured formulations may migrate and render some products, such as food packaging, unacceptable. In most cases, this problem does not exist with EB- cured products. Coatings formed from similar formulations but cured with UV or EB radiation may differ in their physical properties, such as scratch resistance and swelling resistance. This is conceivable, since the two processes are fundamentally different.[4]

Organic molecules become electronically excited or ionized after absorption of energy. For the transformation of organic molecules from the ground state to the excited state, energies typically in the range from 2 to 8 electron-volts (eV) are required.[5] The excited molecules are able to enter into chemical reactions leading to chemically reactive products that initiate the polymerization, cross-linking, and grafting reactions.

Ionization of organic molecules requires higher energy. The ionization process generates positive ions and secondary electrons. When reacting with suitable monomers (e.g., acrylates), positive ions are transformed into free radicals. Secondary electrons lose their excess energy, become thermalized, and add to the monomer. The radical anions formed this way are a further source of radicals capable of inducing a fast transformation.[5]

In industrial irradiation processes, either UV photons with energies between 2.2 and 7.0 eV or accelerated electrons with energies between 100 and 300 eV are used. Fast electrons transfer their energy to the molecules of the reactive substance (liquid or solid) during a series of electrostatic interactions with the outer sphere electrons of the neighboring molecules. This leads to excitation and ionization, and finally to the formation of chemically reactive species.[5] Photons, on the other hand, are absorbed by the chromophoric site of a molecule in a single event. In UV curing applications, special photoinitiators are used that absorb photons and generate radicals or protons. The fast transformation from liquid to solid can occur by free radical or cationic polymerization, which in most cases is combined with cross-linking. In liquid media, the transformation takes typically 1/100 of a second to 1 s. However, in a rigid polymeric matrix, free radicals or cationic species last longer than a few seconds. A post- or dark-cure process proceeds after irradiation, and the result is a solid polymer network.[5]

In summary, UV and electron beam technology improves productivity, speeds up production, lowers cost, and makes new and often better products. At the same time, it uses less energy, drastically reduces polluting emissions, and eliminates flammable and polluting solvents.

The technology is widely used to protect, decorate, or bond items, including fiber optics, compact discs, DVDs, credit cards, packaging, magazine covers, medical devices, automotive parts, aerospace products, and more.

References

1. Berejka, A. J., in *Proceedings 3rd Annual Green Chemistry and Engineering Conference: Moving Toward Industrial Ecology* (June 29–July 1, Washington, DC), p. 35 (1999).
2. Berejka, A. J., *RadTech Report*, 17, No. 5, September/October, p. 47 (2003).
3. Datta, S., Chaki, T. K., and Bhowmick, A. K., in *Advanced Polymer Processing Operations* (Cheremisinoff, N. P., Ed.), Noyes Publications, Westwood, NJ, Chapter 7, p. 158 (1998).
4. Davidson, R. S., *Exploring the Science, Technology, and Applications of U.V. and E.B. Curing*, SITA Technology Ltd., London, p. 27 (1999).
5. Mehnert, R., Pincus, A., Janorsky, I., Stowe, R., and Berejka, A., *UV&EB Curing Technology & Equipment*, John Wiley & Sons, Chichester/SITA Technology Ltd., London, p. 1 (1998).

2

Review of Elements of Radiation Science and Technology

2.1 UV and Visible Radiation

Ultraviolet radiation is part of the electromagnetic spectrum and shows wavelength in the range from 40 to 400 nm. It is known as *nonionizing* or *actinic* radiation. Ultraviolet radiation is divided into the following four regions[1]:

Vacuum UV (VUV): 40–200 nm

UV C: 200–280 nm

UV B: 280–315 nm

UV A: 315–400 nm

Vacuum UV (VUV) is strongly absorbed by quartz, which is used as the envelope material for bulbs. Because of that and because of its small penetration depth, VUV is not suitable for usual radiation curing.

The radiation with wavelengths between 400 and 760 nm is *visible light*, which is useful and even advantageous for certain applications.[2] The energy from visible light is less powerful than that from ultraviolet radiation sources, and consequently, curing times are longer. Another advantage of visible light is that there is less scatter of the longer wavelengths than of those in the UV range, and therefore greater depth of cure can be achieved.[3]

UV energy can be generated by the use of either *electric current* or *microwave energy*. In the first method, the current is passed through a sealed quartz envelope (bulb) having an electrode on each end. It creates an arc across the electrodes, which vaporizes the fill of the bulb (typically mercury) and creates plasma. In the second method, the quartz bulb does not have electrodes and the power is supplied by magnetrons, creating a microwave field around the bulb. Mercury inside the bulb reacts with the microwave energy and also creates *plasma*. In both cases, the plasma operates at about 6,000 K, which generates UV energy.[4]

Plasma is essentially an ionized gas, consisting of a mixture of interacting positive ions, electrons, neutral atoms, or molecules in the ground state or any higher state of any form of excitation as well as photons. Since charge carrier pairs are generated by ionization, the plasma as a whole remains electrically neutral.[5]

Plasma is generated by energy transfer to a system of atoms or molecules until a gas phase is formed and a high degree of ionization is achieved. In the sources of polychromatic UV light, plasma is formed using mercury vapor; the source of energy is electricity, and the discharges in mercury vapor are excited by AC or DC, radio frequency, or microwave energy. Mercury is suitable as the fill gas for this process because of its relatively low ionization potential and sufficient vapor pressure in the range of temperatures used. Moreover, it is chemically inert toward the electrode and wall materials and has a favorable emission spectrum. The excited levels of the mercury atoms are high enough to allow excitation transfer to other metals, such as iron, gallium, lead, cobalt, etc., which allows the modification of the spectral output of the light source.[6]

In industrial applications, photons with energies ranging from 2.2 to 7.0 eV are used. The mercury discharge, as used in mercury arc lamps, produces a polychromatic spectrum with intensive emission lines ranging in energy levels from 2.8 to 6.0 eV.[7]

Monochromatic UV radiation is emitted by excimer lamps. In excimer lamps, microwave discharge[8] or a radio-frequency-driven silent discharge[9] generates excimer excited states of noble gas halide molecules, which decay by the emission of monochromatic UV radiation. Currently, excimer emissions from XeCl with wavelength 308 nm and energy of 4.02 eV and from KrCl with wavelength 222 nm and energy of 5.58 eV reach power levels required for fast cure speeds. Radiation from CuBr excimer with wavelength 535 nm and energy of 2.2 eV allows visible light curing. The emission from Xe_2 excimer with wavelength 172 nm and energy of 7 eV is used for special surface cure applications.[7]

2.1.1 Light Emission from Mercury Gas Discharge

UV light emission from mercury gas discharge with sufficient intensity results from self-sustaining discharges, such as glow and arc discharge. The glow discharge occurring at mercury vapor pressures of about 10^{-2} mbar and current in the range of 0.01–0.1 A (referred to as *abnormal glow*) produces emission of strong 185 and 254 nm radiation. The low-pressure mercury lamps operate in this region.[6] A typical arc discharge occurs at mercury vapor pressures from 0.1 to 10 bar and currents up to 10 A. Medium-pressure mercury lamps operate in the region of arc discharge with the most intensive transition at 365 nm, while various emission lines and some background radiation appear in the UV and visible ranges of the spectrum.[6]

Conventional mercury lamps are essentially sealed quartz tubes with tungsten electrodes at either end and filled with mercury metal and vapor and a starter gas (usually argon). When a high voltage is applied across the electrodes, the starter gas is ionized:

Review of Radiation Science and Technology

$$Ar \rightarrow Ar^+ + e^- \qquad \text{Ionization}$$

$$Ar^+ + e^- \rightarrow Ar^* \qquad \text{Excitation}$$

Ar* Excited state of argon

Recombination of the ionized electron with the argon cation produces an electronically excited argon atom, which can energize and subsequently ionize a mercury atom:

$$Ar^* + Hg \rightarrow Hg^+ + e^-$$

$$Hg^+ + e^- \rightarrow Hg^*$$

Hg* Excited state of mercury

The electronically excited mercury atom generated by the recombination of the mercury cation with an electron loses its energy radiatively. The above are only a few of the processes that take place in the lamp, but the combined effect is the emission of light in the UV and visible regions and the generation of heat.[10] The heat vaporizes some of the mercury metal. The mercury cations are conducting, and the current passing across the electrodes rises until a steady state is reached.

The spectral output of the mercury gas discharge matches the absorption spectra of many photoinitiators well; however, there are regions where more radiant power is required for specific application. An example is pigmented systems, where the pigment absorbs strongly at wavelengths below 360 nm. To generate required transitions, metal halides, such as gallium and iron halides, are added to the mercury discharge. Gallium addition produces strong spectral lines at 403 and 418 nm. Iron addition generates strong multiline emission between 350 and 400 nm.[1] Spectral outputs of some metal halide lamps compared to that of a standard mercury lamp are in Figure 2.1.

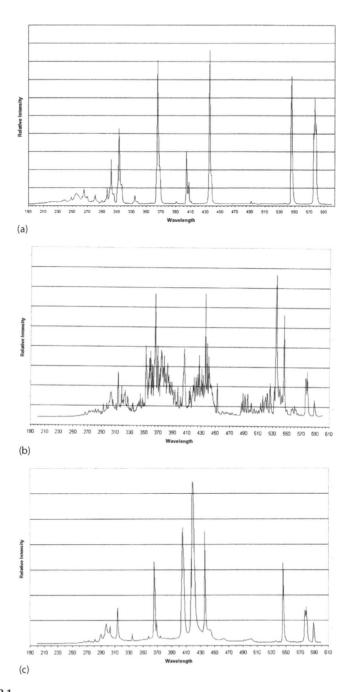

FIGURE 2.1
Spectral outputs of some typical metal halide lamps compared to that of a standard mercury lamp. (Courtesy of American Ultraviolet Company.) (a) mercury barrier discharge lamp; (b) iron additive lamp; and (c) gallium additive lamp.

2.1.2 Light Emission from Microwave-Excited Discharge

Microwave radiation, generated by a magnetron, can be directly transmitted to a quartz tube containing argon as a start gas and a certain amount of liquid mercury. The microwave energy produces a strong electric field inside the bulb, which ionizes argon. At the same time, the mercury is heated up and vaporized and becomes excited and to some extent ionized. Mercury vapor plasma is formed, which can operate at mercury pressures up to 20 bar (290 psi).[1] Spectral outputs of commercial microwave-driven lamps are shown in Figure 2.2.

2.1.3 Generation of Monochromatic UV Radiation

Excimer and nitrogen lasers are sources of pulsed monochromatic UV radiation with a variety of interesting wavelengths and high-output peak powers. However, they cannot compete with mercury arc lamps in large-area, fast-cure applications because of their low pulse repetition rates.[11]

Excimer lamps, a relatively recent new development, represent a new class of monochromatic UV sources.[12] Excimers (excited dimers, trimers) are weakly bound excited states of molecules that do not possess a stable molecular state.[13]

The term stands for an electronically activated molecule in the gas phase consisting of two atoms. The formation process is the same as in lasers[12]:

1. Excitation of atoms in the gas phase within an electrical discharge:

$$A \rightarrow A^*$$

2. Formation of excimer molecules:

$$A^* + B \rightarrow AB^*$$

3. Decomposition of excimer molecules and generation of photons:

$$AB^* \rightarrow A\ B + h\nu$$

where h is the Planck's constant and ν is the frequency of the photon emitted.

The wavelength of the emitted light depends on the choice of atoms A and B. Besides the principle of excimer formation, the technical parameters for the lamp and discharge are responsible for the quasi-monochromatic character of the emitted spectrum.[12] The most important commercial excimers are formed by electronic excitation of molecules of rare gases (He_2, Ne_2, Ar_2, Kr_2, Xe_2), rare gas halides (e.g., ArF, KrF, XeCl, XeF), halogens, and mercury halogen mixtures (HgCl, HgBr, HgI). They are unstable and decay by spontaneous optical emission. The peak maximum is at 308 nm with a half peak full width of about 5 nm.[13] Dielectric barrier "silent" discharges[9] or microwave discharges[8] can be used to produce quasi-stationary or continuous incoherent excimer radiation.

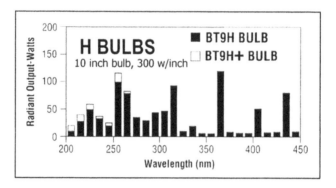

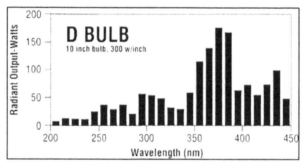

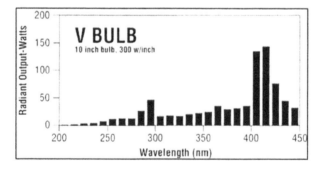

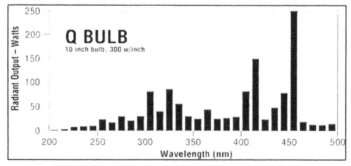

FIGURE 2.2
Spectral output of commercial microwave-driven lamps. (Courtesy of UV Fusion Systems.)

Excimers as radiation sources have the following characteristics[14]:

- They are very efficient energy converters. Theoretically, 40%–50% of the energy deposited by electrons can be converted to radiation.[9,15]
- Due to the absence of a stable ground state, no self-absorption of the excimer radiation can occur.
- In most cases, low-pressure excimer UV sources show a dominant narrow-band (1–3 nm) transition. They can be considered to be quasi-monochromatic.
- Excimer systems can be pumped with very high-power densities. Consequently, extremely bright UV sources can be built.

For industrial applications of excimer UV sources, the dielectric barrier and the microwave discharge are simple, reliable, and efficient excitation modes. There are a large number of vacuum UV (VUV), UV, and visible light transitions available. This allows a selective photoexcitation for many systems.[14] Some sources of monochromatic UV light for industrial applications and their characteristics are shown in Table 2.1.

2.1.4 UV Radiation from UV Light-Emitted Diode

Traditional UV lamps produce UV energy by generating an electric arc inside a chamber filled by an ionizing gas (typically mercury) to excite atoms, which then decay and emit photons, as described in the previous section. The emitted photons cover a broad range of the electromagnetic spectrum, including some infrared and even some visible light. Only about 25% of it is in the safe UV-A range.[16] A UV light-emitting diode (LED) generates UV energy in an entirely different way. As an electric current (electrons) move through a semiconductor device called a diode, it emits energy in the form of photons. The specific materials in the diode determine the wavelengths of these photons and, in the case of UV-LEDs, the input is in a very narrow range, typically ±20 nm. The wavelength is dependent on the band gap between the excited state and the ground state of the semiconductor material used.[15] The electrical to optical conversion efficiency of UV-LEDs is much higher. Moreover, the UV-LED generates considerable less heat.[15] While arc and microwave UV systems are broadband emitters with a range of output between 200 and 445 nm, common wavelength peaks for UV-LED systems are 365, 385, 395, and 405 nm. The emitted wavelengths of UV-LEDs are represented by narrow bell-shaped distributions with a nominal irradiance peak. The exact wavelength where the peak occurs as well as the shape of the distribution is entirely dependent on the structure of the material used, as mentioned above, and it is not adjustable as the diodes are fabricated. The comparison between the outputs of mercury arc and UV-LED systems is shown in Figure 2.3.

TABLE 2.1

Monochromatic UV Sources for Industrial Applications

Source	Emission Generated by	Main Emission Wavelengths, nm	Typical Irradiance W/cm^2	Main Applications	Note
Low-pressure mercury lamp	Low-pressure glow discharge	254 and 185	0.1–1.0	Liquid crystal displays, photoresist technology	Weaker emission lines at 313 and 578 nm
Excimer lamp	Dielectric barrier (silent) discharge	172, 222, 308	1–10	UV curing, pollution control, surface modification	
	Microwave Low pressure discharge	222, 308	1–5		
	High pressure		12–64		
UV LED	As electric current moves through a semiconductor device (diode), the energy is emitted in the form of photons	365, 375, 385, 395, 405	2–10	Printing, coating, adhesives, inkjet marking, bonding and sealing of electronic displays, automotive, wood coatings	Developments focus mainly on creating higher irradiance and new shorter wavelengths.

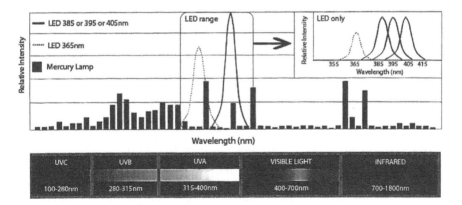

FIGURE 2.3
Difference between outputs of mercury arc lamps and UV-LED lamps. (Courtesy of Phoseon.)

All these differences clearly lead to the necessity to find the proper chemistry in order to achieve the desired results.[16] For example, much of the chemistry for broadband emitters relies heavily on photoinitiators developed for broad spectrum. As a result, not all previously formulated broadband UV chemistry will work with monochromatic LEDs. Therefore, in many cases, the chemistry must be reformulated to react and accomplish the same or similar cure results within the more restrictive but also incredibly more intense band of LED output.[17]

One of the most challenging aspects of UV-LED cure is a poor surface cure due to oxygen inhibition. Oxygen in its ground state has a "diradical" nature and is highly reactive toward radical species.[18] As a result, it scavenges radicals to form less reactive compounds, which can terminate the growing chain via radical-to-radical interaction. The result of oxygen inhibition is decreased rate of polymerization and ultimately compromised performance of the material being cured.[19] A large-scale research has established that the type of photoinitiator is one of the best ways to deal with oxygen inhibition.[20,21]

2.2 Ionizing Radiation Types

As pointed out in Chapter 1, ionizing radiation includes high-energy electrons (electron beam), γ-rays, and X-rays. These not only are capable of converting monomeric and oligomeric liquids into solids, but can also produce major changes in properties of solid polymers. Also, in comparison to UV and visible radiation, they can penetrate considerably deeper into the material.

2.2.1 Electron Beam Energy

In principle, fast electrons are generated in a high vacuum (typically 10^{-6} Torr) by a heated cathode. The electrons emitted from the cathode are then accelerated in an electrostatic field applied between cathode and anode. The acceleration takes place from the cathode that is on a negative high-voltage potential to the grounded vessel as anode. The accelerated electrons may sometimes be focused by an optical system to the window plane of the accelerator.[22]

The energy gain of the electrons is proportional to the accelerating voltage and is expressed in electron volts (eV), which represents the energy gained by a particle of unit charge by passing the potential difference of 1 V. The electrons leave the vacuum chamber only if their energy is high enough to penetrate the 15–20 μm thick titanium window of the accelerator. Details about the equipment used to produce an electron beam are discussed in Chapter 4.

When an electron beam enters a material (this includes the accelerator exit window, the air gap, and the material being irradiated), the energy of the accelerated electrons is greatly altered. They lose their energy and decelerate almost continuously as a result of a large number of interactions, each with only small energy loss.

Electrons, as any other charged particles, transfer their energy to the material, and through that they pass in two types of interaction:

- In collisions with electrons of an atom, resulting in material ionization and excitation.
- In interaction with atomic nuclei, leading to the emission of X-ray photons. This is referred to as *bremsstrahlung* (from German, meaning "breaking radiation,"), an electromagnetic radiation emitted when a charged particle changes its velocity—slows down—due to such interaction.

Somewhat simplified, the process of interaction of high-energy electrons with organic matter can be divided into three primary events:

1. *Ionization:* Ionization takes place only when the transferred energy during the interaction is higher than the bonding energy of the bonding electron:

$$AB \rightarrow AB^+ + e^-$$

At almost the same time, the ionized molecule dissociates into a free radical and a radical ion:

$$AB^+ \rightarrow AB\,{}^\bullet + {}^\bullet B^+$$

2. *Excitation:* Excitation moves the molecule from the ground state to the excited state:

$$AB \rightarrow AB^*$$

The excited molecule eventually dissociates into free radicals:

$$AB^* \rightarrow A^\bullet + B^\bullet$$

3. *Capture of electron:* This process is also an ionization. Electrons with a still lower energy can be captured by molecules. The resulting ion can dissociate into a free radical and a radical ion:

$$AB + e^- \rightarrow AB^-$$

$$AB^- \rightarrow A^\bullet + {}^\bullet B^-$$

Besides these primary reactions, various secondary reactions take place in which ions or excited molecules participate. The final result of these three events is that through the diverse primary and secondary fragmentations, radicals are formed and the complete cascade of reactions triggered by the primary excitation of molecules may take up to several seconds. The energy deposited does not always cause change in the precise position where it was originally deposited, and it can migrate and affect the product yield considerably.

The primary effect of any ionizing radiation is based on its ability to excite and ionize molecules, and this leads to the formation of free radicals, which then initiate reactions such as polymerization and cross-linking or degradation. Accelerated electron beams have energy sufficient to affect the electrons in the atom shell, but not its nucleus, and can therefore only initiate chemical reactions. Typically, the reactions initiated by electron beam are extremely fast and are completed in fractions of a second.

Electrons that are capable of electronically exciting and ionizing organic molecules, such as acrylates, epoxides, etc., must have energies in the range from 5 to 10 eV. Such electrons can be produced from fast electrons by the energy degradation process in solids, liquids, and gases. These secondary electrons show energy distribution with the maximum in the range from 50 to 100 eV. In contrast to fast electrons, exhibiting energies in the keV and MeV range, secondary electrons are capable of penetrating solids and liquids only a few nanometers. Consequently, they generate ions, radicals, and excited molecules in "droplets" along the paths of the fast electrons. Samuel and Magee denoted such droplets containing several pairs of ions, radicals, and excited molecules as "spurs."[23] Chemical processing by electron beam involving polymerization and cross-linking of monomers and oligomers, and cross-linking, modifications, grafting, and degradation of polymers

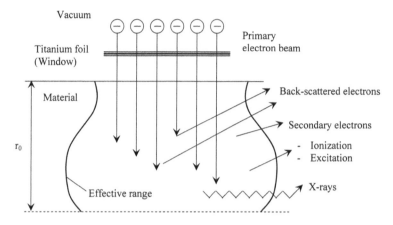

FIGURE 2.4
Process of generation of reactive species by high-energy electrons. (From Garratt, P.G., in *Strahlenhartung*, Curt R., Ed., Vincentz Verlag, Hannover, In German, p. 61, 1996.)

is induced by different reactive species formed initially by electrons in the spurs.[24,25] Figure 2.4 illustrates schematically the process of generation of reactive species.[26]

Radiation processing with electron beam offers several distinct advantages when compared to other radiation sources, particularly γ-rays and X-rays:

- The process is very fast, clean, and can be controlled with a great deal of precision.
- There is no permanent radioactivity because the machine can be switched off.
- In contrast to γ-rays and X-rays, the electron beam can be steered relatively easily, thus allowing irradiation of a variety of physical shapes.
- The electron beam radiation process is practically free of waste products and therefore is no serious environmental hazard.

2.2.2 Gamma Rays

Gamma rays (or γ-rays) represent electromagnetic radiation emitted from excited atomic nuclei of unstable atoms (radionuclides) as an integral part of the process whereby the nucleus rearranges itself into a state of lower excitation (i.e., lower energy content). A gamma ray is a packet of electromagnetic energy—a photon. These photons are the most energetic ones in the electromagnetic spectrum. Essentially, they are emitted by *radioactive decay* and have energies in the range from 10^4 to 10^7 eV. All γ-rays emitted from a given *radioactive isotope* have the same energy. Gamma rays penetrate matter tissues farther than *beta* or *alpha* particles, producing ionization (electron disruption) in their path. In living cells, these disruptions result in damage to the DNA and other cellular structures, eventually causing the death of the organism or rendering it incapable to reproduce. The gamma radiation does not create residuals or impart radioactivity in the materials exposed to it. They are similar to X-rays (see below).

Gamma rays ionize matter by three main processes: the photoelectric effect, Compton scattering, and pair production. In the wide energy range of 100–1 MeV, Compton scattering is the main absorption mechanism, in which an incident γ-photon loses enough energy to eject an electron in an atom of the irradiated matter, and the remainder of its energy is emitted as a new γ-photon with lower energy. Gamma irradiation using ^{60}Co source has a low dose rate, or dose absorbed by the matter per unit time (on the order of 10^{-3} kGy s^{-1}). The dose rate of γ-rays is much lower than that of electron beam. Compared to electron beam, gamma irradiation has a much higher penetration, which is an advantage when irradiating products with large volume. However, the absorbed dose decreases exponentially with the increasing depth of penetration following the Beer-Lambert Law:

$$I_t = I_0 e^{-at} \tag{2.1}$$

where I_t is the intensity of the radiation after passing through the thickness t, I_0 is the initial intensity, and a is the coefficient of linear absorptivity. The attenuation reduces the dose uniformity across the matter.

The most widely used radioactive isotopes in medical and industrial applications are cobalt 60 (^{60}Co), cesium 137 (^{137}Cs), and iridium 192 (^{192}Ir). The half-life of ^{60}Co is 5.3 years, that of ^{137}Cs is 30 years, and the half-life of ^{192}Ir is 74 days.[27] When used for irradiation, the isotope is generally in the form of a pellet size, 1.5×1.5 mm, loaded into a stainless steel capsule and sealed or in the form of "pencils." Unlike electron beam radiation or X-rays, gamma rays cannot be turned off. Once radioactive decay starts, it continues until all the atoms have reached a stable state. The radioisotope can only be shielded to prevent exposure to the radiation. The most common applications of gamma rays are sterilization of single-use medical supplies, elimination of organisms from pharmaceuticals, microbial reduction in and on consumer products, cancer treatment, and processing of polymers (cross-linking, polymerization, degradation, etc.). It should be noted that the products that were irradiated by gamma rays do not become radioactive and therefore can be handled normally.

Strength (or power) of a gamma radiation source is referred to as *radioactivity*.[28] It is defined as spontaneous changes in a nucleus accompanied by the emission of energy from the nucleus as a radiation. Units of radioactivity are *curie* (Ci) or *bequerel* (Bq). Radioactivity is defined as the number disintegrations (dis) of radioactive nuclides per second. Bequerel is the SI unit of radioactivity, which is 1 dis/s. However, this is a very small amount of radioactivity, and traditionally it is measured in units of curie.

1 curie = 3.7×10^{10} Bq. For a gamma radiator, source power may be calculated from the source activity, such that 10^6 curie is approximately 15 kW of power.[28]

2.2.3 X-Rays

X-radiation (also called Röntgen radiation) is a form of electromagnetic radiation. X-rays have a wavelength in the range from 10 to 0.01 nm, corresponding to frequencies ranging from 3×10^{16} to 3×10^{19} and energies in the range of 120 eV–120 keV. They are shorter in wavelength than UV rays. Their name comes from their founders, Wilhelm Conrad Röntgen, who called them X-rays to signify an unknown type of radiation.[29]

X-rays span three decades in wavelength, frequency, and energy. From about 0.12–12 keV they are classified as soft X-rays, and from 12 to 120 keV as hard X-rays, due to their penetrating abilities. X-rays are a form of ionizing radiation, and as such, they can be dangerous to living organisms.

There are two different atomic processes that can produce X-ray photons. One process produces *bremsstrahlung* radiation (see Section 2.2.1), and the other produces K-shell or characteristic emission. Both processes involve a change in the energy state of electrons. X-rays are generated when an electron is accelerated and then made to rapidly decelerate, usually due to interaction with other atomic particles. In an X-ray system, a large amount of electric current is passed through a tungsten filament, which heats the filament to several thousand degrees centigrade to create a source of free electrons. A large potential is established between the filament (the cathode) and a target (the anode). The two electrodes are in a vacuum. The electrical potential between the cathode and anode pulls electrons from the cathode and accelerates them as they are attracted toward the anode, which is usually made of tungsten. X-rays are generated when free electrons give up energy as they interact with the orbital electrons or nucleus of an atom. The interaction of the electrons in the target results in the emission of a continuous *bremsstrahlung* spectrum and also characteristic X-rays from the target material. Thus, the difference between γ-rays and X-rays is that the γ-rays originate in the nucleus and X-rays originate in the electrons outside the nucleus (surrounding it) or are produced in an X-ray generator.[30–33]

X-rays are primarily used for diagnostic radiography in medicine[34] and in crystallography. Other notable uses are X-ray microscopic analysis, X-ray fluorescence as an analytical method, and industrial radiography for inspection of industrial parts, such as tires and welds.

There are reports and patents covering the use of X-rays in processing of a variety of parts made from polymers, particularly in the form of advanced fiber-reinforced composites.[35–39]

A comparison of the three common industrial sources of ionizing radiation is in Table 2.2.

TABLE 2.2

Comparison of Sources of Ionizing Radiation

	Source of Radiation		
Characteristic	Gamma Rays	Electron Beams	X-Rays
Power source	Radioactive isotope[a]	Electricity	Electricity
Power activity	Half-life 5.27 year	Electrical on-off	Electrical on-off
Properties	Photons (1.25 MeV) $\lambda = 1.0 \times 10^{-3}$ nm	Electrons Mass $= 9.1 \times 10^{-31}$ kg	Photons $\lambda = 4.1 \times 10^{-3}$ nm
Charge	None	1.60×10^{-9} coulombs	None
Equipment	Easy to operate and maintain	Complicated to operate and maintain	Complicated to operate and maintain
Emission	Isotropic, cannot be controlled	Unidirectional (can be scanned and bent by magnets)	Forward peaked
Penetration	Exponential attenuation	Finite range, depending on energy	Exponential attenuation
Source attenuation	Continuous attenuation requires regular addition of source	No attenuation	No attenuation
Shielding	Continuous operation requires more shielding	Can be switched on and off, less demanding on shielding	Can be switched on and off, less demanding on shielding
Dose rate	10 kGy/h 2.8×10^{-3} kGy/s	360,000 kGy/h 100 kGy/s	960 kGy/h 0.27 kGy/s

[a] Mainly Cobalt-60.

2.2.4 Other Types of Ionizing Radiation

Subatomic particles other than electrons, such as positrons and protons, can be accelerated to high velocities and energies—usually expressed in terms of center-of-mass energy—by machines that impart energy to the particles in small stages or nudges—ultimately achieving very high-energy particle beams—measured in terms of billions and even trillions of electron volts. Thus, in terms of their scale, particles can be made to perform as powerful missiles for bombarding other particles in a target substance or for colliding with each other as they assume intersecting orbits.

High-energy beams involving the above particles are created in particle accelerators, in which a charged particle is drawn forward by an electro-static field with a charge opposite that of the particle (like charges repel one another, opposites attract); as the particle passes the source of each field, the charge of the field is reversed so that the particle is now pushed onto another field source.

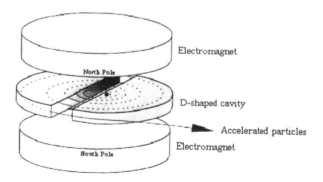

FIGURE 2.5
Schematic of a cyclotron.

An *ion beam* is a type of particle beam consisting of ions. High-energy ion beams are produced like electron beams by particle accelerators, with the difference that a *cyclotron*, another particle accelerator, is used (see Figure 2.5). Cyclotrons accelerate charged particles using a high-frequency, alternating voltage (potential difference). A perpendicular magnetic field causes the particles to spiral almost in a circle so that they reencounter the accelerating voltage many times. Irradiation effects of an ionic beam are different from the ionizing radiation (gamma rays and EB) due to differences in *linear energy transfer* (LET), the average energy deposited in the material used by a projectile particle along its path. In general, the LET of an ion beam is larger than that of gamma rays and electron beam, depending on the particle mass and energy. Applications of ionic beams reported in literature are in modifications of mechanical properties of carbon-fiber-reinforced plastics,[40,41] surface modification of polytetrafluoroethylene,[42] and the production of fuel cells.[43]

2.3 Comparison of UV and EB Processes

In summary, the initiation of polymerization and cross-linking reactions in polymeric systems is different for these two processes. Most of the events initiated by UV radiation, i.e., by photons, occur near the surface because the photon absorption is governed by the Beer-Lambert law (see Section 2.2.2). In contrast, the reactive species produced by electron beam are dispersed randomly throughout the entire thickness of the material (see Figure 2.6).

The penetration depths as well as the amount of energy (i.e., dose) deposited can be regulated with a great deal of accuracy (see Section 6.1).

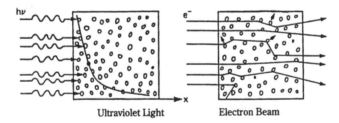

FIGURE 2.6
Comparison of UV and EB radiations (r_0 = substrate thickness).

In spite of these differences, the two processes can be used for the same or similar applications, such as printing, coatings, and adhesives, that is, where only thin layers are being processed. However, for thicker layers, films, and sheets, even opaque and highly filled, as well as composite materials, electron beam radiation is the method of choice.

Other issues regarding the differences between UV and EB processes and applications are discussed in Chapter 1. The recent developments of EB equipment, which represent substantial size and power reductions, enable the EB process to become an alternative to the UV process in several applications, such as graphic arts, where previously only UV equipment was used. A comparison of UV and EB curing processes is in Table 2.3.

The other types of ionizing radiation mentioned above are also used in processing of polymeric systems but not as frequently as electron beams and mainly for specialized applications.

TABLE 2.3

Comparison of UV and EB Curing

Process/Factor	UV Cure	EB Cure
Capital cost	Low	Moderate to high
Energy source	UV lamps	Accelerated electrons
Energy consumption	Moderate to high	Low
Energy activity(curing)	Low	Low to high
Initiation	Photo-induced free radical	Free high-energy electrons
Catalyst	Photoinitiators	None or prorads
Materials	Acrylics and others	Acrylics and others
Inert gas	No (in most cases)	Required
Penetration (typical), μm		
Pigmented	50	400
Clear	130	500
Conversion, %	90	95+

References

1. Mehnert, R., Pincus, A., Janorsky, I., Stowe, R. W., and Berejka, A., *UV&EB Curing Technology and Equipment*, Vol. I, John Wiley & Sons, Chichester/SITA Technology, London, p. 49 (1998).
2. Kosar, J., *Light Sensitive Systems*, John Wiley & Sons, New York (1966).
3. Koleske, J. V., *Radiation Curing of Coatings*, ASTM International, West Conshohocken, PA, p. 30 (2002).
4. Gray, D. E., Ed., *American Institute of Physics Handbook*, McGraw-Hill, New York (1983).
5. Yasuda, H., *Plasma Polymerization*, Academic Press, Orlando, FL (1985).
6. Mehnert, R., Pincus, A., Janorsky, I., Stowe, R. W., and Berejka, A., *UV&EB Curing Technology and Equipment*, Vol. I, John Wiley & Sons, Chichester/SITA Technology, London, p. 45 (1998).
7. Mehnert, R., Pincus, A., Janorsky, I., Stowe, R. W., and Berejka, A., *UV&EB Curing Technology and Equipment*, Vol. I, John Wiley & Sons, Chichester/ SITA Technology, London, p. 12 (1998).
8. Frank, J. D., Cekic, M., and Wood, H. C., *Proceedings of the 32nd Microwave Power Symposium*, Ottawa, Canada, p. 60 (1997).
9. Eliasson, B., and Kogelschatz, U., *Appl. Phys.*, B46, p. 299 (1988).
10. Davidson, R. S., *Exploring the Science, Technology and Applications of U.V. and E.B. Curing*, SITA Technology, London, p. 14 (1999).
11. Mehnert, R., Pincus, A., Janorsky, I., Stowe, R. W., and Berejka, A., *UV&EB Curing Technology and Equipment*, Vol. I, John Wiley & Sons, Chichester/ SITA Technology, London, p. 83 (1998).
12. Roth, A., *RadTech Europe '97, Conference Proceedings* (June 16–19, Lyon, France), p. 92 (1997).
13. Rhodes, Ch. K., *Excimer Lasers, Topics Applied Physics*, Vol. 30, Springer, Berlin (1984).
14. Mehnert, R., Pincus, A., Janorsky, I., Stowe, R. W., and Berejka, A., *UV&EB Curing Technology and Equipment*, Vol. I, John Wiley & Sons, Chichester/SITA Technology, London, p. 93 (1998).
15. Eckstrom, D. J., et al., *J. Appl. Phys.*, 64, p. 1691 (1988).
16. Kiyoi, E., The State of UV LED Curing, An Investigation of Chemistry and Applictions, *UV-LED Curing Technology*, RadTech International, NA, p. 60 (2016). ebook #2.
17. Heathcote, J., UV-LED Overview, Part 1, Operation and Measurement, *UV-LED Curing Technology*, RadTech International, NA, p. 62 (2016). ebook #2.
18. Crivello, J., and Dietliker, K., *Photoinitiators for Free Radical, Cationic & Anionic Photopolymerization*, 2nd ed., John Wiley & Sons, New York, p. 259 (1998).
19. Desauer, R., *Photochemistry, History and Commercial Applications of Hexaarylbiiidazoles, All about HABI*, Elsevier, Oxford (2006).
20. Arcenaux, J., "Mitigation of Oxygen Inhibition in UV-LED, UVA, and Low-Intensity UV Cure," *UV+EB Technology*, Vol. 1, No. 3 (2015).
21. Wyrostek, M., and Salvi, M., "Photoinitiator Selection for LED – Cured Coatings," *UV+EB Technology*, Vol. 3, No. 2, p. 12 (2017).
22. Samuel, A. H., and Magee, J. L., *J. Chem. Phys.*, 21, p. 1080 (1953).

23. Clegg, D.W., in *Irradiation Effects on Polymers* (Clegg, D. W., and Collyer, A. A., Eds.), Elsevier, London (1991).
24. Singh, A. and Silverman, J., in *Radiation Processing of Polymers* (Singh, A., and Silverman, J., Eds.), Carl Hanser Verlag, Munich, p. 1 (1992).
25. Mehnert, R., in *Ullmann's Encyclopedia of Industrial Chemistry*, Vol. A22 (Elvers, B., Hawkins, S., Russey, W., and Schulz, G., Eds.), VCH, Weinheim, p. 471 (1993).
26. Garratt, P. G., in *Strahlenhartung*, (Curt R., Ed.), Vincentz Verlag, Hannover, p. 61 (1996). (in German.)
27. *Radioactive Isotopes Used in Medical and Industrial Applications*, NDT Resource Center, www.ndt-ed.org (June 7, 2009).
28. Drobny, J.G., *Ionizing Radiation and Polymers, Principles, Technology and Applications*, Elsevier, Norwich, NY, p. 4 (2013).
29. Kevles, B. H., *Naked to the Bone Medical Imaging in the Twentieth Century*, Rutgers University Press, Camden, NJ, p. 19 (1996).
30. Dendy, P. P., and Heaton, B., *Physics for Diagnostic Radiology*, CRC Press, Boca Raton, FL, p. 12 (1999).
31. Feynman, R., Leighton, R., and Sands, M., *Feynman Lectures on Physics*, Vol. 1, Addison-Wesley, Reading, MA, p. 2 (1963).
32. L'Annunziata, M., and Baradei, M., *Handbook of Radioactivity Analysis*, Academic Press, New York, p. 58 (2003).
33. Grupen, C., Cowan, G., Eidelman, S. D., and Stroh, T., *Astroparticle Physics*, Springer, Berlin, p. 109 (2005).
34. Smith, M. A., Lundahl, B., and Strain, P., *Med. Device Technol.*, 16, No. 3, p. 16 (2003).
35. Ramos, M. A., Catalao, M. M., Schacht, E., Mondalaers, W., and Gil, M. H., *Macromol. Chem. Phys.*, 203, No. 10–11, p. 1370 (2002).
36. Wang, C.-H., et al., *J. Phys. D*, 41, p. 8 (2008).
37. Sanders, C. B., et al., *Radiat. Phys. Chem.*, 46, p. 991 (1995).
38. Berejka, A. J., *RadTech Report*, 16, No. 2, March/April, p. 33 (2002).
39. Galloway, R. A., Berejka, A. J., Gregoire, O., and Clelland, M. R., U.S. Patent Application 20080196829, August 21 (2008).
40. Kudoh, H., Sasuga, T., and Seguchi, *Radiat. Phys. Chem.*, 48, p. 545 (1996).
41. Seguchi, T., et al., *Nucl. Instrum. Meth. Phys. Res. B*, 151, p. 154 (1999).
42. Choi, Y. J., Kim, M. S., and Noh, I., *Surf. Coat. Technol.*, 201, p. 5724 (2007).
43. Yamaki, T., et al., in *Proceedings of Conference Proton Exchange Fuel Cells* (October 29–November 3, Cancun, Mexico) (2006).

3

Ultraviolet Curing Equipment (with Ruben Rivera)

The main applications of UV curing lamps and systems are surface curing of inks, coatings, and adhesives. They typically were operating originally in the 200–450 nm wavelength range, with lamp power as high as 240 W/cm (600 W/in.) of lamp length.[1] However, as of this writing, the typical lamp power has increased for a wide range of applications up to 700 W/cm (1750 W/in.). In some cases, the equipment may operate in wavelengths up to 750 nm wavelength, which is already in the visible light range.

The UV curing equipment consists essentially of the following three components:

1. *Lamp (or bulb)*: The electrical energy supplied to the bulb is converted into UV energy inside the bulb.

2. *Lamp housing*: The housing is designed to direct and deliver to the substrate or the part to be irradiated. The lamp housing contains the reflector that focuses the ultraviolet energy generated by the lamp. This is also referred to as the irradiator.

3. *Power supply*: The power supply delivers the energy needed to operate the UV lamp.

A typical UV curing unit may house one or more lamps. Most frequently, the material to be cured is passed underneath one or more lamps using a moving belt. The speed of the belt determines how long the web is exposed to the UV. The UV generated by the lamp is reflected by a reflector, which may either focus or defocus it, depending on the process.

The lamp performance, which affects cure, is expressed by the following characteristics[2]:

1. *UV Irradiance* is the radiant power arriving at a surface per unit area. The most commonly used unit of irradiance is W/cm^2. It varies with lamp output power, efficiency, the focus of its reflector system, and its distance to the surface. The intense peak of focused power directly under the lamp is referred to as *peak irradiance*. Although not a defined term, *intensity* is occasionally used to refer to irradiance. Irradiance incorporates all of the contributing factors of electrical power,

efficiency, radiant output, reflectance, focus, bulb size, and lamp geometry. Higher irradiance at the surface will produce correspondingly higher light energy within the coating. The depth of cure is affected more by irradiance than by the length of exposure. The effects of irradiance are more important for higher absorbance (more opaque) films. Higher irradiance can permit the use of less photoinitiator.[2]

2. *UV exposure* is the radiant energy per unit area arriving at a surface. UV exposure represents the total quantity of photons arriving at a surface in contrast to irradiance, which is the rate of arrival. Exposure is inversely proportional to speed under any given light source, and proportional to the number of exposures (in multiple rows of lamps it is the number of "passes"), and is the time integral of irradiance with the usual units J/cm^2 and mJ/cm^2. It is simply the accumulated energy to which the surface was exposed.

3. *Infrared radiance* is the amount of infrared energy primarily emitted by the quartz envelope of the UV source. This energy is collected and focused with the UV energy on the surface of the substrate to the extent depending on the IR reflectivity and efficiency of the reflector. Infrared radiation can be expressed in exposure or irradiance units, but most commonly, the surface temperature it generates is of primary interest. The heat it produces may be a benefit or a nuisance.[2]

4. *Spectral distribution* describes the relative radiant energy as a function of the wavelength emitted by a bulb or the wavelength distribution of energy arriving at a surface, and it is often expressed in relative (normalized) terms or as absolute figures to compare various lamps and light sources. It can be displayed as a plot or a table, and in this form it allows comparison of various bulbs and can be applied to calculations of spectral power and energy.[2]

3.1 UV Lamps

There are a large number of lamps used to initiate the reactions in the processed material. They differ in design and source of radiation. The list of commercially available lamps[3] are as follows:

- Mercury lamps (low-, medium-, and high-pressure)
- Electrodeless (microwave-powered) lamps
- Excimer lamps
- Xenon lamps (free running and pulsed)

- Spot cure lamps
- Continuous wave (CW) and pulse lasers
- Light-emitting diodes

While any of the above sources find specific applications, medium-pressure mercury (microwave and arc) and xenon lamps are currently most widely used in industrial applications, and the use of light-emitting diodes is growing rapidly and it appears that at one point will replace mercury arc lamps completely.

3.1.1 Medium-Pressure Mercury Lamps

The most frequently used lamp for UV curing processes is a medium-pressure mercury lamp. Its emission spectrum can be used to excite the commonly used photoinitiators. Moreover, this type of lamp has a relatively simple design, is inexpensive, can be easily retrofitted to a production line, and is available in lengths up to 8 ft. (2.5 m). Power levels in common use are in the range of 40–240 W/cm, and even higher levels are available for special applications.[3]

A medium-pressure mercury arc lamp consists of a sealed cylindrical quartz tube with tungsten electrodes on each end. The quartz tube is chosen because of its high transmission of ultraviolet radiation and a low coefficient of thermal expansion, which enables it to withstand the high operating temperatures. The tube is press-sealed at the ends and running through the seals are thin strips of molybdenum foil to which the electrodes are connected. The tube contains a specific small amount of mercury and an inert starter gas (usually argon). Some manufacturers also add a small amount of krypton-85 gas that aids easier starting and arc maintenance. An example of such a lamp is in Figure 3.1.[4]

The spectral output of the medium-pressure lamp can be altered by "doping," i.e., adding a small amount of metal halide to the fill material, with mercury. Commonly doped lamps are iron and gallium lamps[5] but can also

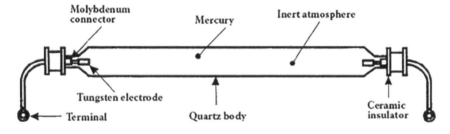

FIGURE 3.1
Medium-pressure mercury lamp. (Adapted from *UV&EB Technology*, Vol. I, SITA Technology Ltd., UK, 1998, p. 53.)

include antimony, magnesium, cobalt, lead, and thallium. Medium-pressure lamps have long lifetimes (typically in excess of 3,000 h), although the intensity of the emitted light and the relative intensity of the spectral lines change with time. These lamps will typically lose 10% of their initial output within the first 100–200 h of operation. The lifetime is also affected by the switching cycle (turning the lamp on and off) as well as by the input power. This affects their performance to the point that the lamp might strike, but the output is so low that it does not trigger the reaction. Therefore, it is necessary to check its output periodically by a spectroradiometer.

3.1.2 Electrodeless Mercury Lamps

Electrodeless lamps have emission spectra similar to those of medium-pressure doped and undoped lamps (the manufacturers often use the term *additive lamps*). They are actually medium-pressure mercury lamps, but they differ from arc lamps in that the tubes do not have electrodes (see Figure 3.2). The excitation of mercury atoms is by microwave power generated by magnetrons. The system containing the power modules (magnetrons) and reflectors is integral, and because of that, it is somewhat larger than the equivalent medium-pressure arc lamp system. This has to be taken into consideration when designing or retrofitting a line. The currently available width of the lamp is up to 26 cm (10 in.). The modules are designed so that when wider webs are processed, several modules are mounted end to end to cover any width, with virtually no interruption to the continuous line of focused UV across the web.

In the current design of an electrodeless microwave-powered lamp, the microwave energy is generated by two 1.5 or 3.0 kW magnetrons with a frequency of 2,450 MHz and fed through an antenna system to each waveguide. The microwave energy coupled with precisely designed slots is uniformly distributed over the length of the bulb. Magnetrons and waveguides are cooled by filtered air, which is also passed into the microwave chamber through small holes in the reflector. This downward airflow cools the lamp and keeps the lamp and the reflector clean.

In general, the electrodeless lamp contains a small amount of mercury and a neutral starter gas. The mercury pressures during its operation are in the

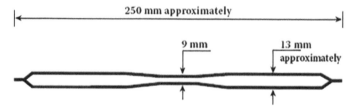

FIGURE 3.2
Electrodeless lamp. (Courtesy of Hanovia.)

TABLE 3.1

Characteristics of Electrodeless Lamps

Lamp System	Bulb Type	Bulb Diameter, mm	Peak Irradiance W/cm^2 (UVA)
F-450	H	9	2.5
(350 W/in.)	D	9	6.0
	V	9	2.0
F-450	D	13	5.1
(350 W/in.)	D	11	5.6
	D	9	6.0
F-600	H	13	2.7
(560 W/in.)	D	13	8.0
	V	13	2.4
F-600	D	13	8.0
(560 W/in.)	D	11	9.6
	D	9	9.9

Source: Courtesy Fusion UV Systems.

5–20 bar range, which is higher than for arc lamps, with pressures only in the 1–2 bar range. The characteristics of the lamp can be modified by the addition of metal halides in controlled quantities to the gas. The frequently used D or V lamps contain iron or gallium salts, respectively.[6] Some characteristics of microwave-powered lamps are shown in Table 3.1.

A variety of gas fills are available, and these cover a fairly large range of UV and visible spectra. The bulbs are interchangeable and can be changed easily, so it is easy to use the proper lamp for a specific job. The advantages of the electrodeless lamp systems are:

- The modular lamp-reflector system is simple and easy to install and change.
- The bulb does not decrease in output or in its spectral distribution during its entire life.
- They have longer lifetimes than arc lamps (4,000–6,000 h vs. 1,000–1,500 h).
- They have short warm-up times (2–3 s).
- After being switched off, the lamps do not have to cool down before reigniting.

Disadvantages of electrodeless lamps are mainly associated with the cost of the equipment maintenance, since the magnetrons have to be replaced every about 5,000–8,000 h. Additional shortcomings are higher cost of power due to lower efficiency of the microwave- generating process. In a multiple-lamp system, multiple bulbs and magnetrons are replaced, and the cost is a trade-off of the longer operating life and stability of exposure.

3.1.3 Low-Pressure Mercury Lamps

Low-pressure mercury lamps consist of either a quartz cylinder with electrodes on both ends or doped soda lime glass also known as soft glass. Although most lamps are cylindrical, other shapes are also produced.[7] Inside the lamp is a mixture of mercury and argon at a pressure of 10^{-2}–10^{-3} Torr. The emission from this lamp is 254 nm, and with high-quality quartz, some light with 189 nm wavelength is produced. Low-pressure lamps are of low power and therefore are not used for the cure of coatings, but are well suitable for applications where slow cure rate is tolerated, such as liquid crystal displays or in resist technology for the production of microchips.

3.1.4 High-Pressure Mercury Lamps

High-pressure mercury lamps operate at pressures of about 10 atm, and there are essentially two types used in industrial applications. The *point source lamp* focuses on a small-diameter spot, thus delivering an intense radiation to that spot. The *capillary lamp* is used for narrow webs up to approximately 20 cm (8 in.) wide.[4] They are capable of producing a wider and more continuous spectrum than the medium-pressure lamps and operate with higher power (150–2,880 W/cm). Their disadvantage is a relatively short operating life, typically hundreds of hours.

3.1.5 Excimer Lamps

Excimer lamps are quasi-monochromatic light sources available in UV wavelengths. The light is produced by silent electrical discharge through gas in the gap between two concentric quartz tubes. Electronically activated molecules are produced in the gas phase and decompose within nanoseconds to produce photons of high selectivity. This process is similar to the process in excimer lasers.

The most important difference between excimer and standard UV lamps is that the former are incoherent radiation sources and can therefore be used for large area applications. Different wavelengths can be produced by choosing different gas fills in the gap between the quartz tubes.

The wavelength of the emitted light depends on the choice of atoms. Besides the principle of excimer formation, the technical parameters for the lamp and discharge are responsible for the quasi-monochromatic character of the emitted spectrum. The most important commercial excimers are formed by electronic excitation of molecules of rare gases (He_2, Ne_2, Ar_2, Kr_2, and Xe_2), rare gas halides (e.g., ArF, KrF, XeCl, and XeF), halogens, and mercury halogen mixtures (HgCl, HgBr, and HgI). They are unstable and decay by spontaneous optical emission. The lamp is completely free of ozone.[8] Currently, excimer lamps with wavelengths of 308, 222, and 172 nm are commercially available.

Stimulated (laser) excimer emission can be generated in pulsed high-pressure glow discharges. Dielectric barrier (silent) discharges[9] or microwave discharges[10] can be used to produce quasi-stationary or continuous incoherent excimer radiation.

With the narrow range of wavelength of the excimer lamp, a specific initiator can be selected with the maximum efficiency in that range. Thus, it is possible to reduce the amount of photoinitiator from that used for conventional UV sources and reduce the cost of the material and amounts of unreacted initiator in the finished coated substrate.[11] Excimer lamps can be used for both free radical and cationic photoinitiators.[12]

Currently, there are two types of commercial excimer lamps: barrier-discharge- and microwave-driven lamps.

3.1.5.1 Barrier-Discharge-Driven Excimer Lamps

The excimer lamp of this type is made by Heraeus Noblelight (Hanau, Germany) and consists of two concentric cylindrical quartz tubes, fused together on both ends. The excimer gas mixture is filled in the gap between the two tubes and sealed. The electrodes are placed outside the discharge gap (Figure 3.3). They transfer a high-frequency alternating voltage through a barrier discharge into the gas, where the excimers are formed. This technology allows the lamps to work in both continuous and pulsating modes.

IR radiation is not produced, and therefore the lamps are relatively cold, with the surface temperatures being in the range of 35°C–40°C (95°F–104°F).[12] This makes them suitable for heat-sensitive substrates.

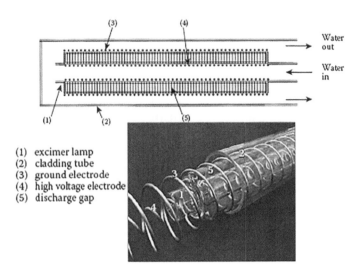

(1) excimer lamp
(2) cladding tube
(3) ground electrode
(4) high voltage electrode
(5) discharge gap

FIGURE 3.3
Barrier discharge excimer lamp. (Courtesy of Heraeus Noblelight.)

In addition to the low surface temperature, excimer lamps have the advantage that they start immediately, so they can be switched on and off as needed. There is no need for standby function and for a shutter system, so no movable parts are necessary on the radiation head. Both continuous and pulse mode operation are possible. The lamps are very compact, which is particularly important for the printing industry, because it is possible to retrofit them into the existing process.

The excimer curing system of this design consists of the following components[8]:

- Bulb of different radiation length in an optimized compact housing
- Power supply
- Cooling unit (using deionized water)
- Inertization (optional)

These lamps are radio frequency (RF) powered and can be easily controlled and switched. The power supply includes all electrical controls, interlocks, and the water-cooling supply system.[13] A complete system is shown schematically in Figure 3.4.

The nature of this design of excimer UV lamps makes them particularly effective in flexo and offset printing. The advantages cited in these applications[8] are:

- No heating of the substrate or press.
- Improved print quality, since there is no pixel displacement due to film distortion.

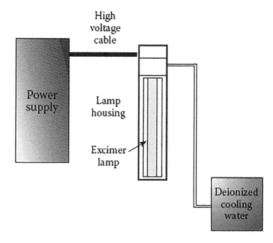

FIGURE 3.4
UV curing system with a barrier discharge excimer lamp. (Courtesy of Heraeus Noblelight.)

- Reduced smell of inks (the lamp operates in a nitrogen atmosphere) and substrates (due to low operating temperature).
- Reduced press downtime, since the lamp does not need warm-up after being switched on or reignited.
- Compact excimer curing system that can therefore be easily retrofitted into existing equipment.

One interesting application of barrier charge excimer lamps is surface matting of acrylate coatings with a wavelength of 172 nm and subsequent UV or electron beam (EB) cure of the entire thickness of the film. The advantage is that no matting additives are needed, and the line speeds are much higher than those in the traditional process[8] (see Section 7.1.5).

Due to relatively low UV radiant power of these types of lamps, nitrogen inerting has to be used,[14] which increases the operating cost.

3.1.5.2 Microwave-Driven Excimer Lamps

Fusion Systems (Gaithersburg, Maryland) at one time offered 240 W/cm microwave-driven excimer lamps within its VIP (Versatile Irradiance Platform) series. They were among the most powerful sources of UV radiant power[15] available in XeCl (peak wavelength 308 nm) and CuBr (peak wavelength 535 nm). These lamps are no longer commercially available.

3.1.6 Xenon Lamps

Xenon lamps are available mainly as tubular and point source bulbs. The radiation produced by this type of lamp is not particularly rich at wavelengths below 400 nm; therefore, their applications are somewhat limited. However, it is possible to pulse xenon lamps, which enables them to achieve high peak irradiances. Commercially available pulsed xenon lamps are available with emissions in the UV and visible spectral range. Alternating the gas fill can produce output rich in UV.

The advantages of pulsing xenon lamps are:

- The pulses are short; consequently, there is no significant heat buildup.
- There is no need to increase temperature to evaporate mercury; thus, xenon lamps operate at considerably lower temperatures.
- The lamps can be turned off completely in microseconds, and thus there is no continuous infrared radiation and no shutters have to be installed.
- There is no need for nitrogen inerting.
- With its high peak power, pulsed light penetrates opaque materials more effectively than does continuous light.

Because of the cooling period between pulses, these lamps operate at much lower temperatures and are suitable for manufacturing processes where the product does not tolerate high temperatures.

Xenon lamps do not generate ozone and thus can be used in processes where the materials are sensitive to ozone. An additional advantage is that there is no need for plant ventilation systems for the removal of ozone.

Pulsed curing systems are widely used in the manufacture of medical devices, electronics, semiconductors, and optical fibers. Pulsed xenon lamps can be made in a variety of shapes to fit specific requirements, such as 360° illumination. Examples of different designs are shown in Figures 3.5 and 3.6.

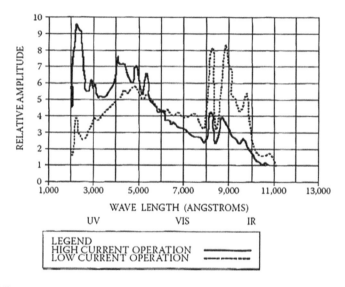

FIGURE 3.5
Spectral dependence of a xenon lamp on electrical operating parameters. (Courtesy of Xenon Corporation.)

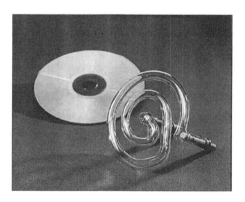

FIGURE 3.6
Spiral xenon lamp used for DVD bonding. (Courtesy of Xenon Corporation.)

3.1.7 Light-Emitting Diode Lamps

In principle, a light-emitting diode is a two-lead semiconductor light source. It is a p-n junction diode that emits light when activated. When a suitable voltage is applied to the leads, electrons are able to recombine with electron holes within the device, releasing energy in the form of *photons*. The key property of a diode is that it conducts electricity in only one direction. A p-n junction (positive-negative junction) is made of many layers of semiconductive materials where each layer is less than 1 μm thick.[16]

The UV-LED chips, the building blocks for UV-LED lamps, are made, as mentioned above, from very tiny, thin slices of semiconductive materials such as gallium nitride (GaN) or aluminum gallium nitride (AlGaN) thin layers (chips) that are doped or impregnated with additives. The type of material and its composition determine the emitted wavelength. Commercial ultraviolet light-emitting diode (UV-LED) curing lamps are semiconductor light sources that emit very discrete wavelengths of light energy, resulting in a single, narrow, bell-shaped emission spectrum as shown in Section 2.1.4. They generate UV-curing light using an array of surface-mounted light-emitting diodes instead of traditional metal halide or mercury bulbs. The benefit of an array is that if one diode fails, the light intensity of the surface light is only minimally affected. LED lamps emit only in a narrow band of UV wavelength producing a single peak, centered at a specific wavelength, reducing heat transfer to the substrate. Mercury lamps emit in a broad spectrum with many sharp peaks including visible and infrared (IR) wavelengths. IR wavelengths transfer heat directly to the material and substrate, potentially damaging them. In contrast to mercury lamps, even the low-energy ones, LED lamps are far more energy efficient. Schematic of a basic UV-LED lamp is shown in Figure 3.7.

The most common wavelengths of the UV-LED lamps are in the UVA region or near the UVA region. These typical wavelengths are 365, 385, 395, and 405 nm. The LED lamps that emit in this region have a very tight peak with as little as 5–7 nm on either side of the peak wavelength. When a

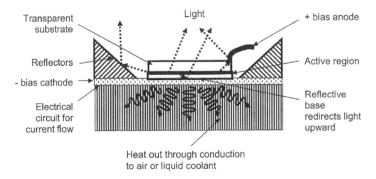

FIGURE 3.7
Schematic of the UV-LED lamp. (From Sahara, R., "UV LED Lens Technology," *Proceedings of RadTech 2008 Conference*, Chicago, May 4–7, 2008.)

mercury lamp is compared with an LED lamp, the relative intensity shows a much higher peak irradiance in the so-called *curing wavelengths* (as mentioned above). UV-LED lamps can use optical concepts such as micro optics and micro reflectors. The LEDs typically emit their energy in mainly in 120 degree angle. Spherical lenses above chips allow emission in a mainly in 60 degree angle. These lenses can be made mainly from borosilicate or quartz.[17] These are considered primary optics, and while they have a tighter beam angle, they are still not collimating the UV. This can be done with secondary optics that focus light at a given distance.[18] In addition, LED lamps are safe, while mercury lamps are a risk in the workplace. If the outer shield of an LED array is damaged, there are no harmful effects. If the outer bulb of a mercury lamp breaks, then intense UV radiation is emitted. UV exposure can cause eye and skin burns and other discomforts.[19] Properly designed UV-LEDs have an operating life span of over 40,000 h. Mercury arc lamps have a life span at most of 1,500 h. Because of the design of UV-LED lamps, they allow an instant on and off switching, which is impossible with mercury lamps. The differences between typical mercury arc lamps and UV-LED lamps are listed in Table 3.2. Figure 3.8 depicts the emitted wavelength curves for commonly used UV light-emitting diodes.

TABLE 3.2

Comparison of Typical Mercury Arc Lamps and UV-LED Lamps

Property	Mercury Arc Lamps	UV-LED Lamps
Life span, hours	1,500–2,000	Up to 40,000
On/Off	10 minutes	Instant
Temperature generated, °C	~350	~60
Output consistency	Drops up to 50% over time	95% +
Maintenance	Bulb replacement and reflector cleaning	Minimal
Input power	Large	Small
Environmental	Mercury waste	Mercury-free
	Generate ozone	Ozone-free
Curing energy efficiency	—	Saves 50%–75%

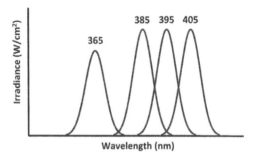

FIGURE 3.8
Emitted wavelength curves for common UV-LEDs.

Peak irradiance, also called intensity, is the radiant power arriving at a surface per-unit area. With UV curing, the surface is the cure surface of the substrate or part, and a square centimeter is the unit area. Irradiance is expressed in units of watts or milliwatts per square centimeter (W/cm² or mW/cm²). Peak irradiance is instrumental in penetration and aiding surface cure. Peak irradiance is affected by the output of the engineered light source, the use of reflectors or optics to concentrate or contain the rays in a tighter surface impact area, and the distance of the source from the cure surface. The irradiance for UV-LEDs at the cure surface decreases quickly as the distance between the source and the cure surface increases.

Energy density, also called dose or radiant energy density, is the energy arriving at a surface per-unit-area during a defined period of time (dwell or exposure). A square centimeter is again the unit area and radiant energy density as expressed in units of joules or millijoules per square centimeter (J/cm² or mJ/cm²). Energy density is the integral of irradiance over time or the area under the bell curve. A sufficient amount of energy density is necessary for full cure.[20]

Because of the differences between mercury arc lamps and UV-LED lamps, different photoinitiator types and combinations than those used in standard mercury curable formulations are often required. In most cases, formulations for mercury arc lamps are not compatible with UV-LED, while UV-LED some formulations would cure under both mercury and UV-LEDs. This unique characteristic is driving formulators to move to dual cure chemistries (R. Rivera, personal communication, October 2019).

The effect of chemistry is also a driver in the push for lower wavelength UV-LED lamps in the UVC region. Free radical formulations are susceptible to curing with a tacky surface when exposed to air (oxygen inhibition) (R. Rivera, personal communication, October 2019). Mercury lamps produce many different wavelengths in the UV region, such as UVA, UVB, and UVC. Since UV-LED lamps are nearly monochromatic (see above), there is no UVC emitted that could aid in the oxygen inhibition. There are several solutions to get around this surface cure problem. The first is that the formulation can be modified. The next is that a nitrogen inerting system can be added, although this may not be practical due to size/space limitations as well as cost. Lastly, the UVC region can be used by either UV mercury or UV-LED lamps (R. Rivera, personal communication, October 2019).

The UVC chips are still relatively low-power lamps compared to UVA LED systems and are typically measured in mW/cm² instead of W/cm². The other issues with these products are their relatively short lifetimes of 3,000–10,000 h and high costs, which are associated with the low yield rates in producing these deep UV devices (R. Rivera, personal communication, October 2019).[16] Recent research results have shown that the surface cure challenges that have previously prevented adoption of full UV-LED curing solutions are being overcome.[21] When paired with a UVA LED system, providing just a little UVC exposure for postcure not only provides a tack-free surface, but also reduces the total dose requirements. When viable UVC implementations are then combined with the advancements in formulations, further improvements can be seen to minimize the dosage needed while still achieving a hard surface cure.[21]

In spite of some of the issues mentioned above, the adoption of UV-LEDs as replacement of mercury arc lamp technologies in industrial practice is accelerating due to their following advantages:

- Energy efficiency
- Lack of heat generated by LED lamps
- Higher power of LED lamps
- Longer lifetimes of LED lamps
- Instant on/off
- Increased regulatory pressure on mercury[19]

3.2 Lamp Housings

The lamp housing of mercury arc lamps has the function to direct and deliver the energy to the substrate or piece being cured. The reflector system contained in the housing reflects the energy generated by the lamp and the way it depends on its design. The older, parabolic reflectors just reflected the energy in a flood pattern. Semielliptical reflectors direct the energy to a focal point. The newest type, elliptical reflectors with a greater wrap around the lamp, delivers a more focused energy distribution forming a narrow curing zone with an increased amount of energy at the far focal point. With the development of 3D processes and the many surfaces of a piece that need to be exposed to UV, semielliptical reflectors will likely be used more frequently. It is quite possible that new reflector designs will soon appear.[22] The two basic reflector designs are depicted in Figure 3.9.

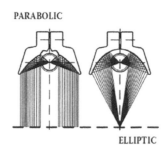

FIGURE 3.9
Two basic designs of reflectors. (Courtesy of IST America.)

The material that will reflect the entire range of UV wavelengths is *pure* aluminum. UV reflectors are made from highly polished (specular) sheet (mill) aluminum with vapor-deposited aluminum on the surface with a protective inorganic overcoat. Reflector supports, which provide the proper curvature of the reflector, are made from extruded aluminum or castings in different lengths and shapes, and they are also designed to provide lamp cooling. Arc lamp electrode seals must be kept below 250°C (482°F); the lamp jacket must be kept under 600°C–800°C (1,110°F–1,470°F); and aluminum starts to deteriorate above 250°C (482°F). Air-cooling or water-cooling systems are used to help cool the lamp seals and quartz jacket.[23] By contrast, microwave-powered bulbs are maintained at a nearly constant temperature over their entire length, reducing stress on the quartz and simultaneously reducing the generation of ozone.

Only a perfectly smooth surface exhibits directed (specular) reflection, i.e., where the angle of reflectance equals the angle of incidence. This, however, is not the case under real-life conditions, where the surface micro-roughness cannot be avoided, and consequently, the UV can be slightly diffused over a finite range of reflection angles.[23]

UV reflectors also reflect radiation energy in the visible and infrared (IR) energy spectral regions at least as efficiently as in the UV region.[22,23] This is undesirable in such cases where the substrate or the piece being irradiated is heat sensitive. To overcome this, the surface of the reflector is coated with special dielectric coatings ("cold mirror"), which, when applied in multiple layers, will reflect UV but transmit visible and IR energy. An example of a cold mirror reflector is shown in Figure 3.10.

Often, in order to contain an inert atmosphere in the curing zone, a quartz plate or window is placed between the lamp and the substrate. The filter must be cooled sufficiently to prevent it from becoming an IR radiation source itself.[22] The disadvantage of this design is that part of the UV transmission is lost, and the plate also interferes with the air cooling of the substrate. A few designs use water-cooled reflectors, but this only cools the

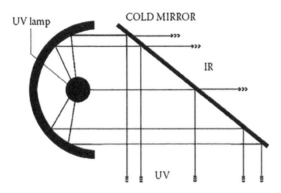

FIGURE 3.10
Example of a cold mirror reflector. (Courtesy of Honle UV America.)

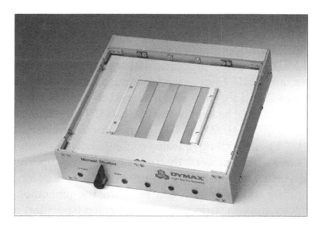

FIGURE 3.11
Manual shutter open. (Courtesy of Dymax Light Curing Systems.)

lamp and does not alter the IR or visible energy reflected to the substrate. In some designs, two quartz cooling tubes are placed between the lamp and the substrate. Deionized water circulates through the tubes. Up to 50% of the infrared emission from the lamp can be absorbed by this type of filter,[24] although there is also a significant reduction in the UV. An additional control of IR energy is to coat the filter with a reflective coating ("hot mirror"), which reflects a limited amount of the IR radiation back to the reflectors without significantly affecting the transmitted UV.[22,24] For this to function properly, the reflector in the lamp housing must absorb the reflected IR.

In practical applications, some lamps, notably medium-pressure mercury arc lamps, cannot be started or restarted instantaneously and it needs some time for restrike. Interruption of lamp action would be totally unacceptable in continuous production. Moreover, the heat from burning lamps could cause deterioration of the substrate if the line is in standby mode (stopped). Therefore, shutters (see Figure 3.11) are installed that close over the lamp, which is immediately switched to half power. Shutters can be incorporated in the reflector design, in a clamshell fashion, or may rotate to enclose the lamp.[25] Shutters are not normally used with microwave-powered lamps, as they are virtually instant on and off.

3.3 Power Supply and Controls

3.3.1 Power Supply Systems

Until recently, a typical power supply for mercury arc lamps would only operate at discreet power levels, some going from standby to high, others supplying medium power. This is sufficient for most curing processes but limits the capability of the user to control a highly specialized process.

This type of power supply often limits the process speed, which in turn means that the energy usage could be much more than required for the given chemical curing system. When operating at high power, the by-products of generating UV energy, such as heat and visible light, are at their most intense.[22]

Medium-pressure mercury arc lamps are normally powered by AC. Because of the high impedance of the plasma arc in the lamp, the maintenance of plasma requires a high electrode potential. The required potential gradients are up to 30 V per centimeter of length, and the open-circuit voltage should be up to twice the operating voltage. This means that main voltage is insufficient for practically all lamps. Therefore, in order to ignite and operate the lamp, a step-up transformer is required. Other important factors entering the design of the power supply are[26]:

- To ignite and operate the lamp, a step-up transformer is required.
- Due to negative voltage-current characteristics of the lamp discharge, a current limiting power supply must be used.
- The burn-in characteristics of the arc lamps have to be taken into account.

Immediately after ignition, mercury is still liquid and the discharge takes place in the fill gas (argon). At this point, the voltage is low and the current is limited essentially by the short-circuit current delivered by the power supply. As the temperature within the lamp increases, mercury vaporizes. The lamp impedance increases, and this causes the lamp voltage to increase and the current to decrease. After about 1 min, the burn-in period is finished and the lamp reaches stationary conditions.

The most common type of power supplies for mercury lamps with large arc lengths combine a step-up transformer and a "ballast" capacitor placed in the secondary circuit of the transformer (see Figure 3.12). Since the lamp

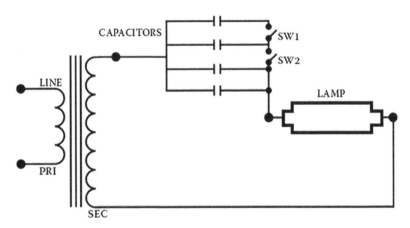

FIGURE 3.12
Power supply for mercury lamps.

power depends on its capacitance, the stepwise power can be simply varied by switching capacitors as long as the lamp is operated at appropriate cooling conditions.

Electrodeless medium-pressure mercury lamps are powered by microwave energy. The system contains either one or two microwave generators, two waveguides, and a closed nonresonant microwave chamber formed by a semielliptical metallic reflector with flat ends and a fine metal mesh.[27] Microwave energy is generated by one or two 1.5–3.0 kW, 2,450 MHz magnetrons, and fed through an antenna system to each waveguide. The metallic reflector and the metal screen closure mesh form a microwave chamber, which, from the electrical point of view, has a noncavity structure. The microwave energy coupled into the chamber by precisely designed slots is uniformly distributed over the lamp tube. Standing waves, which would result in resonant distribution of microwave energy, are avoided by slot design and choosing pairs of magnetrons slightly differing in frequency in lamps using two magnetrons. Magnetrons and waveguides are cooled by filtered air, which is also passed into the microwave chamber through small holes in the reflector. This downward airflow also cools the bulb and keeps the bulb and reflector clean.[28] The power supply also contains filament heating transformers, a control board with status, and fault indicators and numerous safety and production interlocks.

3.3.2 Control Systems

Currently, there are several control systems for UV power supplies of mercury arc lamps. One design is based on the use of a transductor as inductive ballast and current control at the same time. The current flowing through the lamp is controlled by premagnetization of the transductor. This allows a continuous lamp power control within the range from 40%–100%. Besides the continuous current control, smooth lamp burn-in, a short dark period between the AC half-waves, early reignition after each voltage half cycle, and other benefits result.[29] Another means of achieving stepless control of current in the lamp is by using thyristors. When compared to transductor control, the dark period is not constant but increases at lower levels of power.[29]

In general, with the above controls, the lamp power depends on the fluctuations of the main voltage. This may cause fluctuations in irradiance and potentially reduce the quality of the product. This can be avoided by using a closed-loop control circuit.[30]

In a closed-loop control circuit, a UV sensor is used to monitor the radiant power of the lamp. In a control unit, the signal is compared with a preset signal or with a signal provided by a tachometer. The resulting difference in the signal is magnified and used to control the lamp current by transductor or thyristor switches.[30] An example of a control system is in Figure 3.13.

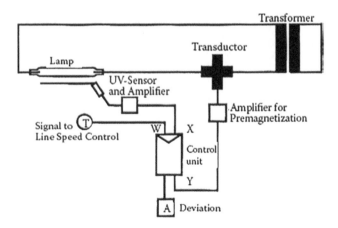

FIGURE 3.13
Example of a control system for a mercury lamp. (Adapted from *UV&EB Technology*, Vol. I, SITA Technology Ltd., UK, 1998, p. 66.)

Recently, electronic control systems for power supplies have been developed. Their reported advantages over conventional systems are[31]:

- Current or power controlled system, independent of main voltage
- Electrical efficiency over 92%
- No external electrical compensation necessary
- Very low adjustable standby power (<40%)
- Large dimming range (15%–110%)
- Compact dimensions and lower weight

This type of control system is operated at medium frequency and is currently limited to a maximum power of about 5 kW.[30]

3.4 UV-LED Curing Systems

UV-LED lamp systems consist of number of components, which taken together define the performance of the system. The main components are typically[32]:

1. **Light-emitting diode** (LED) is a solid-state component that generates UV light when subjected to electric current due to the process called electroluminescence. A single UV diode (die) is in Figure 3.14.
2. **Array** is a grouping of LEDs to maximize UV output to achieve the desired curing rate. An array may contain a large number of

FIGURE 3.14
Single diode (chip). (Courtesy of Phoseon.)

FIGURE 3.15
Example of a UV-LED array. (Courtesy of Kopp Glass.)

individual diodes to achieve a desired performance. An example of an array is in Figure 3.15.

3. **Thermal cooling** is a properly designed thermal management system for the removal of heat generated by the LED array. This component ensures low operating temperature and long life span of the lamps. Currently, the arrays are cooled either with air or liquid.

4. **Optics** has to be designed so that it reflects and guides the UV-LED light in such a way that the maximum light reaches the processed media. The design has to balance the fact that LEDs are a "flood" type of light, unlike focused mercury lamps where the light is captured by a reflector and directed to a specific-point focal length. The optical system can be as simple as a quartz window (flood), or a more complex lens or reflector assembly.

Power supply units for UV-LED systems are commonly solid-state devices where the main electrical supply power is typically anywhere between 100 and 240 V AC at either 50 or 60 Hz. Solid-state power supplies automatically accommodate for the given voltage.[33]

Most applications require UV-LED curing systems that consist of more than one LED or LED array in order to achieve not only the desired through-put but to meet the width of the processed media, which can be 1–2 m (40–80 in.).[32]

UV-LED irradiator is UV curing assembly, which includes multiple LED chips or modules, a thermal heat sink, a cooling fan or tube fittings, a mani-fold block, an emitting window, a sheet metal or plastic outer housing, and sometimes, the driver boards.[32]

The irradiance (W/cm^2) produced by UV-LED sources has increased con-sistently over the past few years as the result of advances in diode and lamp technology, and currently their effective outputs are higher than those offered by traditional UV curing lamp technologies. Typically, current UV-LED lamps offer peak irradiance up to 24 W/cm^2 (water cooled) and 16 W/cm^2 (air cooled) at 395 nm, and that number is expected to increase continuously. UV-LED systems at 365 nm are available with peak irradiance of 12 W/cm^2 (water cooled) and 8 W/cm^2 (air cooled).[34] They are being used commercially to cure inks, coatings, and adhesives.[33] References 33 and 34 provide many important details on the subject.

3.4.1 Selected UV-LED Industrial Lamps/Systems

There are many lamps and systems available, dozens of them are supplied by a number of different manufacturers and suppliers. To list them all would take a great deal of space and could become confusing for the reader. For that reason, only several systems are listed below. These are well-established and well-performing lamps and systems from established manufacturers.

1. *X Series™* by **AMS Spectral UV** (Figure 3.16) is suitable for sheet-fed offset and web lithography; flexography; label printing; digital printing; corrugated and plastic printing; panel and screen curing; and wood curing.

   Specifications:

   Wavelengths: 340, 365, 385, 395, 405 nm, multiwave (blended) options available.

   Peak irradiance: up to 32 W/cm^2

   Widths: can be built as wide as 3.2 m (126 in.)

2. *SCi Power and Control System* by **AMS Spectral UV** (Figure 3.17) for X Series™ designs.

3. *Omnicure®* by **Excelitas** (Figure 3.18), UV-LED spot cure system.

4. *Semray® 4103 UV-LED system* by **Heraeus Noblelight** (Figure 3.19). Air-cooled plug-and-play system, peak wavelengths, nm: 365, 385, 395, 405; Maximum irradiance values, W/cm^2: 13 at 365 nm, 15 at 385 nm, 18 at 395 nm, and 17 at 405 nm. Emission window size: 77 × 45 mm.

FIGURE 3.16
X Series™ by AMS Spectral UV.

FIGURE 3.17
SCi Power and Control System by AMS Spectral UV for X Series™ designs.

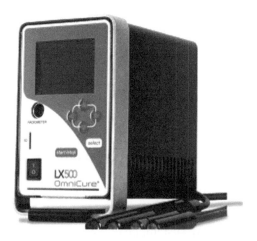

FIGURE 3.18
Omnicure® by Excelitas, UV-LED spot cure system.

FIGURE 3.19
Semray® 4103 UV-LED system by Heraeus Noblelight.

FIGURE 3.20
IRIS UV-LED 250 Mark II by Heraeus Noblelight.

Dimensions on the UV4103 UV4103 Segment Model (W × D × H) in mm: 77 × 136 × 253.

5. *Iris UV-LED 250 Mark II* by **Heraeus Noblelight** (Figure 3.20). Peak wavelengths, nm: 365, 385, 395. Maximum irradiance values, W/cm²: 15 at 365 nm, 22 at 385 nm, 22 at 395 nm. Emission window size: 254 × 26.5 mm. Standard dimensions in mm: 275 × 130 × 50.

Application areas include printing, automotive, aerospace, packaging, electronics, wood coatings, and medical.

6. *Semray UVPC 6003 UV curing system* by **Heraeus Noblelight** (Figure 3.21). The system is cooled by internal fans, has peak wavelength of 395 nm, irradiance at target 65–70 W/m², uniformity >80% at target, and power output range is 40%–100%. The system is used in optical fiber applications.

7. *FireEdge™ FE300* by **Phoseon** (Figure 3.22). Air-cooled product, peak irradiance 3 W/cm² (for 365 nm), 5 W/cm² (for 385/395/405 nm),

FIGURE 3.21
Semray UVPC 6003 UV curing system by Heraeus Noblelight.

FIGURE 3.22
FireEdge™, FE300, by Phoseon.

FIGURE 3.23
FirePower™ FP601 by Phoseon.

small form factor, for pinning or full cure, used for adhesives, coatings, and inks.

8. *Fire Power™ FP 601* (Figure 3.23). Water-cooled product, peak irradiance, 5 W/cm^2 (for 385/395/405 nm), high irradiance, and used mostly for flexo applications.

References

1. Berejka, A. J., in *Conference Proceedings, 3rd Annual Green Chemistry and Engineering* (June 29–July 1, Washington, DC), p. 35 (1999).
2. Stowe, R. W., *RadTech Europe '97, Conference Proceedings*, (June 16–18, Lyon, France), p. 603 (1997).
3. Davidson, R. S., *Exploring the Science, Technology and Applications of UV and EB Curing*, SITA Technology Ltd., London, p. 13 (1999).
4. Knight, R. E., in *Chemistry and Technology of UV and EB Formulations for Coatings, Inks and Paints*, Vol. 1 (Oldring, P. K. T., Ed.), SITA Technology Ltd., London, p. 159 (1991).
5. Davidson, R. E., *Exploring the Science, Technology and Applications of UV and EB Curing*, SITA Technology Ltd., London, p. 16 (1999).
6. Mehnert, R., Pincus, A., Janorsky, I., Stowe, R., and Berejka, A., *UV&EB Curing Technology & Equipment*, John Wiley & Sons, Chichester/SITA Technology Ltd., London, p. 73 (1998).
7. Davidson, R. E., *Exploring the Science, Technology and Applications of UV and EB Curing*, SITA Technology Ltd., London, p. 17 (1999).
8. Roth, A., *RadTech Europe '97, Conference Proceedings* (June 16–18, Lyon, France), p. 91 (1997).
9. Eliasson, B., and Kogelschatz, U., *Appl. Phys.*, B46, p. 299 (1998).
10. Kumagai, H., and Obara, M., *Appl. Phys. Lett.*, 54, p. 2619 (1989).
11. Skinner, D., *RadTech Europe '97, Conference Proceedings*, (June 16–18, Lyon, France), p. 133 (1997).
12. Roth, A., *RadTech Report*, 10, No. 5, September/October, p. 21 (1996).
13. Mehnert, R., Pincus, A., Janorsky, I., Stowe, R., and Berejka, A., *UV&EB Curing Technology & Equipment*, John Wiley & Sons, Chichester/SITA Technology Ltd., London, p. 95 (1998).
14. Mehnert, R., Pincus, A., Janorsky, I., Stowe, R., and Berejka, A., *UV&EB Curing Technology & Equipment*, John Wiley & Sons, Chichester/SITA Technology Ltd., London, p. 102 (1998).
15. Mehnert, R., Pincus, A., Janorsky, I., Stowe, R., and Berejka, A., *UV&EB Curing Technology & Equipment*, John Wiley & Sons, Chichester/SITA Technology Ltd., London, p. 101 (1998).
16. Heathcote, J. "UV-LED Overview, Part I-Operation and Measurement," *UV-LED*, RadTech International, www.radtech.org (2013). ebook, #1.
17. Galbraith, J., "Current State of UV LED Technology," Kopp Glass, koppglass.com (2016).
18. Verhoff, T., Grade, J., and Stahl, A., Optical Concept for High Power UV LED Lamps, in *Conference Proceedings, RadTech Europe Conference* (October 15–16, Munich) (2019).
19. "The Mercury Ban and its effect on the UV LED Printing Industry" *Technical Note*, Phoseon™ Technology, www.phoseon.com (2019).
20. "Irradiance or Energy Density," Technical White Paper, Phoseon Technology, April 2019. www.phoseon.com (2019).
21. Kay, M., Improving Surface Cure with UVC LEDs, *UV+EB Technology*, Vol. 4, No. 1, p. 26 (2018).

22. Whittle, S., *RadTech Report*, 14, No. 1, January/February, p. 13 (2000).
23. Mehnert, R., Pincus, A., Janorsky, I., Stowe, R., and Berejka, A., *UV&EB Curing Technology & Equipment*, John Wiley & Sons, Chichester/SITA Technology Ltd., London, p. 57 (1998).
24. Mehnert, R., Pincus, A., Janorsky, I., Stowe, R., and Berejka, A., *UV&EB Curing Technology & Equipment*, John Wiley & Sons, Chichester/SITA Technology Ltd., London, p. 59 (1998).
25. Mehnert, R., Pincus, A., Janorsky, I., Stowe, R., and Berejka, A., *UV&EB Curing Technology & Equipment*, John Wiley & Sons, Chichester/SITA Technology Ltd., London, p. 63 (1998).
26. Mehnert, R., Pincus, A., Janorsky, I., Stowe, R., and Berejka, A., *UV&EB Curing Technology & Equipment*, John Wiley & Sons, Chichester/SITA Technology Ltd., London, p. 65 (1998).
27. Mehnert, R., Pincus, A., Janorsky, I., Stowe, R., and Berejka, A., *UV&EB Curing Technology & Equipment*, John Wiley & Sons, Chichester/SITA Technology Ltd., London, p. 71 (1998).
28. Mehnert, R., Pincus, A., Janorsky, I., Stowe, R., and Berejka, A., *UV&EB Curing Technology & Equipment*, John Wiley & Sons, Chichester/SITA Technology Ltd., London, p. 72 (1998).
29. Mehnert, R., Pincus, A., Janorsky, I., Stowe, R., and Berejka, A., *UV&EB Curing Technology & Equipment*, John Wiley & Sons, Chichester/SITA Technology Ltd., London, p. 67 (1998).
30. Mehnert, R., Pincus, A., Janorsky, I., Stowe, R., and Berejka, A., *UV&EB Curing Technology & Equipment*, John Wiley & Sons, Chichester/SITA Technology Ltd., London, p. 68 (1998).
31. Beying, A., in *RadTech Europe '97, Conference Proceedings* (June 16–18, Lyon, France), p. 77 (1997).
32. Jennings, S., Larson, B., and Taggard, C., "UV-LED Curing Systems: Not Created Equal," *RadTech eBook #1*, p. 38, RadTech International, www.radtech.org (2013).
33. Heathcote J., "UV-LED Overview Part II Curing Systems," *RadTech ebook #1*, p. 30, RadTech International, www.radtech.org (2013).
34. Higgins, M., "Understanding Ultraviolet LED Wavelength," *UV+EB Technology*, Vol. 2, No. 2, p. 47 (2016).

4

Electron Beam Curing Equipment

Commercial electron beam processors available today vary in accelerating voltage range from about 100 kV up to about 10 MV. The original development of electron beam technology in the 1950s concentrated on machines with accelerating voltages in the range from 1 to 2 MV with beam powers of 5–10 kW, suitable primarily for the cross-linking of plastic materials. During the 1960s, the successful applications required even higher voltages, up to 3 and 4 MV, with power output levels up to 100 kW.[1] Lower-voltage machines in the range below 300 kV were developed in the 1970s. They were designed without a beam scanner, which was used in the higher-voltage equipment. The electron beam in the lower-voltage range was delivered as a continuous "shower" or "curtain" across the entire width of the product. The main applications of the lower-voltage machines were coatings and other thin-film layers.[1] The most recent designs use voltages in the range of 70–125 kV, with electron energies suitable for the processing of thin layers of inks and lacquers less than 25 μm thick.

The principle of producing high-energy electrons is very simple. The electrons are emitted in a vacuum by a heated cathode and accelerated in the electrostatic field applied between cathode and anode. Acceleration takes place from the cathode, which is connected to a negative high-voltage potential, to the grounded accelerator window as anode. Usually, an electron optical system is used to focus the accelerated electrons to the accelerator window plane.

The energy gain of the electrons is proportional to the accelerating voltage. It is expressed in electron volts (eV), i.e., the energy that a particle of unit charge gains by passing a potential difference of 1 V. The electrons leave the vacuum chamber and reach the process zone if their energy is high enough to penetrate the 6–14 μm thick titanium window foils used in the lower-energy range and 40–50 μm thick foils used in the higher-energy range.

As pointed out in Section 2.2.1, stopping the high-energy electrons by a material generates X-rays. For that reason, the electron accelerator and process zones have to be shielded to protect the operator. For electrons with energies up to 300 keV, a self-shielding with lead cladding up to about 1 in. thick is sufficient. For systems where electrons with higher energy are generated, the equipment is shielded by a concrete or steel vault built around it, as shown in Figure 4.1.

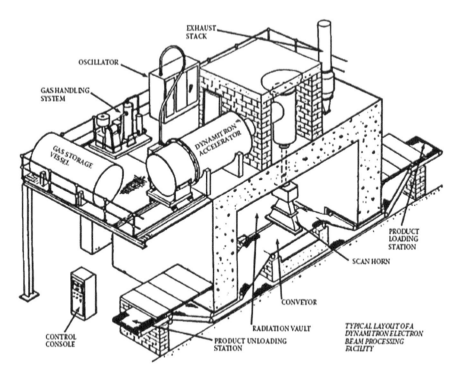

FIGURE 4.1
Traditional design of a high-energy EB unit. (Courtesy of IBA.)

The basic electrical parameters of an electron beam processor are its *acceleration voltage*, the *electron beam current*, and the *electron beam power*. The ratio of electron beam power and the input electrical power defines the efficiency of an electron accelerator.[2] The acceleration voltage determines the energy of the electrons, as pointed out in Section 2.2.1.

The absorption properties of the accelerated electrons in the processed materials are the *absorbed dose*, the *depth dose profile*, the *penetration range*, and the *dose rate*.

Process parameters involve line speed; if dose rate and line speed are combined, the dose delivered to the product to be cured can be calculated. A processor-specific yield factor depends on the relationship between the beam current and line speed. The *dose-speed capacity* of a processor is given by the product of the line speed and the delivered dose at maximum electron beam power.[3]

The abovementioned parameters are defined below: *Acceleration voltage:* Potential difference between cathode and anode of the accelerator, usually expressed in kV or MV and obtained using a resistive voltage divider chain in the high-voltage unit.

Electron beam current: Number of electrons per second emitted from the cathode, measured in mA (1 mA = 6.25 × 10^{15} electrons per second) at the high-voltage unit.

Electron beam power: Product of the acceleration voltage and the electron beam current, expressed in kW (1 kW = 10 mA × 100 kV or 10 mA × 0.1 MV).

Absorbed dose: Mean value of energy of the ionizing radiation absorbed by the unit of mass of the processed material. The SI unit of absorbed dose is 1 Gray (Gy) = 1 J kg^{-1} and a larger, more practical unit is 1 kGy = 10^3 J kg^{-1}. The older unit, used officially until 1986 and still used occasionally in literature, is 1 megarad (Mrad), which is equivalent to 10 kGy.

Depth dose profile: The energy deposition produced by electrons of a given energy in an absorber can be expressed as a function of depth, density, atomic number, and atomic weight of the absorber. The energy loss dE/dx of the fast electrons (expressed in MeV cm^{-1}) is given as a function of the mass per unit of area (g cm^{-2}), which is equivalent to the depth in cm times the density in g cm^{-3}. It is directly proportional to the absorbed dose: $1/\rho \, dE/dx$ = dose × unit area, with dimensions J $g^{-1}cm^2$. In realistic electron beam processes, energy losses occur in the titanium window foil, in the air gap between foil and the coating, and in the substrate.

Dose rate: Dose per unit time, expressed in Gy s^1 = J $kg^{-1}s^{-1}$. At a constant accelerating voltage, it is proportional to the electron beam current.

Line speed: Speed of the material being irradiated, usually expressed in m/min or ft/min. It determines the exposure time of the material.

Delivered dose: The ratio of the dose rate and the line speed:

$$\text{Delivered dose} = \text{Dose rate/Line speed}$$

At a given dose rate, the line speed can be adjusted to obtain the desired dose. On the other hand, the beam current can be controlled by the line speed to maintain a constant delivered dose at all line speeds.

Yield factor: Yield factor is used to characterize the curing performance of an electron processor. It is a constant that relates the delivered dose to the beam current and line speed:

$$\text{Delivered dose} = \text{Yield factor} \times \text{Beam current/Line speed}$$

The unit of yield factor is kGy × m/min/mA.

Dose-speed capacity: Product of the delivered dose and the line speed at the maximum beam power:

$$\text{Dose} - \text{Speed capacity} = \text{Delivered dose} \times \text{Line speed}$$

In usual EB operation, the dose-speed capacity is measured in m/min at 10 kGy. It is the most convenient unit to relate the curing performance of an electron beam processor to the desired process parameters, such as dose and line speed.[4]

The industry defines low-energy machines with accelerating voltage less than 300 kV. The accelerated electrons are generated in a vacuum chamber, they pass through a titanium window foil. Given current window foil technology, the practical low-energy end of the spectrum today is about 80 kV. This is enough for penetrating materials up to about 10 g/m² (10 μm with a density of 1.0), while 300 kV will penetrate about 450 g/m² (450 μm with a density of 1.0). As to the width, the vast majority of low-energy EB systems are used in web processing, with the low end being about 270 mm (10 in.)—typical for the seal tube EB lamps (see Section 4.1.4) up to a maximum of 3,000 mm (120 in.).

Another very important characteristic is the throughput that determines the cost of the operation. The units of the throughput are the product of the dose and line speed, as pointed out above. A typical dose for cross-linking is roughly 100 kGy, so at 150 m/min (500 ft/min) the required throughput of the system would be 15,000 kGy·m/min. For curing of inks coatings, the typical dose is 30 kGy and speed is 400 m/min (1300 ft/min); the required system throughput would then be15,000 kGy·m/min.

4.1 Particle Accelerators

Particle accelerators produce electrically charged particles with speeds slightly over 6,000 miles/s to speeds approaching the speed of light. Accelerated particles are generally formed into a beam and directed at a material called a beam window. The window is thin enough so that the particle beam may pass through it without much energy loss. As this beam passes through the window material, some of the particles are deflected and may interact with atomic nuclei. The final effect depends on the energy and nature of the incident particle. A particle accelerator is depicted schematically in Figure 4.2.[5]

There are several designs of accelerators for industrial use. They are mainly used to accelerate electrons, although they can be adapted to accelerate ions as well. For the purpose of this publication, the accelerators of electrons will be discussed at some length.

Essentially, there are two types of electron accelerators used in industrial electron beam applications. *Direct accelerators*, also referred to as *potential drop accelerators*, require the generation of high electrical potentials equal to the final energies of the accelerated electrons. *Indirect accelerators* produce high electron energies by repetitive application of time-varying electromagnetic

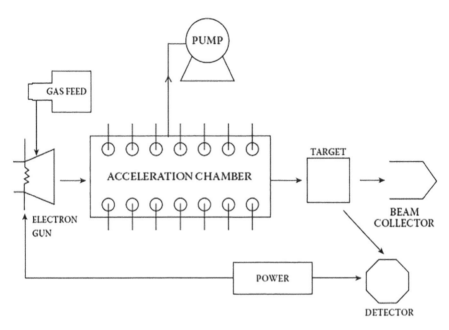

FIGURE 4.2
Schematic of a particle accelerator.

fields (Chrusciel, J., private communication). Direct accelerators are widely used for low- and medium-energy applications because they are able to provide continuous electron beams with high average current and power ratings, which translate to high processing rates. For higher energies (about 5 MeV), indirect accelerators using microwave radiation, very high frequency (VHF) radiation, or pulsed electrical power are considered to be more suitable. Modern accelerators can generate beam powers up to several hundred kilowatts and electron energies up to 15 MeV (Chrusciel, J., private communication).

4.1.1 Direct Accelerators

In principle, a direct electron accelerator consists of a high-voltage generator connected to an evacuated acceleration system. The different direct accelerators currently used employ similar methods for electron emission, acceleration, and dispersion; the differences are in the design of their voltage generators.

The source of electrons is a thermionic cathode, which is almost always a tungsten or thoriated tungsten wire, although tantalum wire or lithium hexaboride pellets are also used as cathodes (Chrusciel, J., private communication). Electron emission is most frequently controlled by the variation of temperature of the cathode, but this can also be done by a grid with variable voltage. The beam of electrons is extracted, focused, and accelerated by an internal electric field, determined by a series of electrodes or dynodes with intermediate potentials obtained from a resistive voltage divider (Chrusciel, J., private communication). The electrons gain energy continuously throughout the length of the acceleration tube. After acceleration, the concentrated electron beam is dispersed by scanning with a time-varying magnetic field with a sweep rate of at least 100 Hz. The divergent beam so formed expands within an evacuated scan horn and then emerges into the air through a thin metallic foil. This foil is usually made from titanium, but other metals, such as aluminum alloys or titanium alloys, are used.[27] In order to minimize the loss of the electron energy, the thickness of a titanium foil is kept in the range between 25 and 50 pm for electron energies above 0.5 MeV, and between 6 and 15 pm for energies below 0.3 MeV. The windows of the scanning horn may be as wide as 2 m (80 in.) (Chrusciel, J., private communication).

4.1.1.1 Electrostatic Generators

Electrostatic generators are based on the principle that high electric potentials can be produced by mechanically moving static charges between low-voltage and high-voltage terminals. The various designs have used endless rubber belts, chains, or metallic electrodes with insulating links, and rotating glass cylinders. Compressed gasses—for example, nitrogen, sulfur hexafluoride (SF_6), Freon, carbon dioxide, and others—are used for high-voltage insulation to reduce the size of the equipment. One example of such a device is a Van de Graaff generator (Figure 4.3), a belt-driven machine, originally developed in the 1930s for research in nuclear physics. Industrial versions capable of delivering several kW of electron beam power with energies up to 4 MeV were developed in the 1950s. They were used for electron beam cross-linking of plastic films, tubing, and insulated wire, and biomedical applications.[6] Today, these machines are no longer competitive with modern high-power machines and are therefore seldom used in industrial applications.

4.1.1.2 Resonant Transformers

Pulsed electron beams produced by these devices are generally limited to an average current of about 5–6 mA because they operate at a relatively low frequency of 180 Hz. The voltage of such an accelerator is not rectified as it is in other industrial accelerators, and the high voltages (typically 1.0 and 2.0 MV) of commercial resonant transformers produce pulsating current at their resonant frequency. This pulsating current makes it difficult to achieve a uniform dose in the irradiated material. These machines consist of a pressure tank, in which the iron-free resonant transformer and the

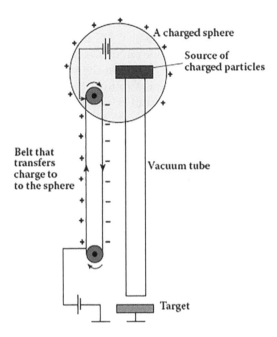

A charged sphere

Source of
charged particles

Belt that
transfers
charge to
to the sphere

Vacuum tube

Target

FIGURE 4.3
Schematic of the Van de Graaff accelerator.

discharge tube are placed. Secondary windings connected in series feed the high-voltage terminal. The system provides a beam only during the negative half cycle, with a voltage variation between zero and the peak value the machine was designed to deliver. Sulfur hexafluoride gas is used for electrical insulation.[7]

4.1.1.3 Iron Core Transformers

Low-frequency transformers with iron cores connected to rectifier circuits can generate potentials up to approximately 1 MV. Traditionally, iron core transformers are insulated by oil, but the recent models use sulfur hexafluoride. The energy rating of these generators is in the range from 0.3 and 1 MeV with beam powers up to 100 kW.[8,9]

4.1.1.4 Insulating Core Transformers

Insulating core transformers (ICTs) were developed in the late 1950s and early 1960s to replace Van de Graaff electrostatic generators to provide more irradiation power for higher processing rates. An ICT consists of a three-phase transformer with multiple secondary windings that are energized serially by iron core segments. These segments are separated by thin sheets of dielectric material. Low-voltage, low-frequency AC power is converted into high-voltage DC power by using rectifier circuits connected to the nearest core segment. This way, the electrical stresses in the system are minimized. ICT designs are capable of higher-voltage ratings than conventional iron core transformers.[8] The insulating medium is sulfur hexafluoride gas, and the line-to-beam power efficiency is close to 85%.[10–13] The terminal voltage of the ICT power supply is connected to an accelerator tube. The electron beam for industrial applications starts from a tungsten filament and is accelerated within the evacuated tube as in other DC accelerators.

ICT machines produced recently have energy ratings from 0.3 to 3.0 MeV and beam power capabilities up to 100 kW. Nearly in 180 of these machines, the majority of them rated for less than 1 MeV, have been installed as of the early 1990s. They are used mainly for cross-linking of heat-shrinkable film, plastic tubing, and electric wire.[8]

4.1.1.5 Cockcroft-Walton Generators

This is essentially a cascade generator. In this type of electron accelerator, the high voltage is produced by the incremental movement of electric charge. The high-voltage system is a condenser-coupled, cascade rectifier, which converts low-voltage, medium-frequency (3 kHz) AC power into high-voltage DC power. The capacitive coupling circuits are connected in series. Sulfur hexafluoride is used as insulation.[14] Cockcroft-Walton generators are made in different designs. Three-phase rectifier circuits are used for low-energy, high-current applications, conventional single-phase cascades for energies between 1 and 3 MeV, and balanced two-phase systems for energies between 1 and 5 MeV with beam power ratings up to 100 kW. Electrical efficiencies are about 75%.[15]

4.1.1.6 Dynamitrons

In the Dynamitron type of accelerator (Figure 4.4), the high-voltage DC power is generated by means of a cascade rectifier circuit energized by high-frequency (100 kHz) AC power. The rectifiers are driven in parallel by a pair of large semicylindrical electrodes that surround the high-voltage column. This arrangement enhances the reliability at the maximum voltage by eliminating the large, high-voltage condensers used in a series-coupled Cockcroft-Walton generator. Because of power loss in the high-frequency oscillator, employing a triode vacuum tube, the electrical efficiency of a Dynamitron accelerator is about 60%.[16,17] These machines have been made with energy ratings from 0.5 to 5 MeV and beam power ratings up to 250 kW. More than 150 Dynamitrons, the majority of which have energy ratings above 1 MeV, are used for the cross-linking of polymeric materials.[8] Some of the larger machines with energy ratings above 3 MeV are also used for sterilization

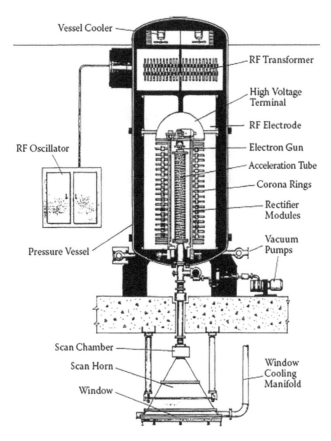

FIGURE 4.4
Schematic of a direct accelerator (Dynamitron) accelerator.

of medical products by energetic electrons.[18] The largest model, which can deliver 250 kW of beam power at energies in the range from 3 to 5 MeV, is equipped with a high-power target for X-ray processing.[19,20]

4.1.2 Indirect Accelerators

This type of accelerator produces high-energy electrons by injecting short pulses of low-energy electrons into a copper waveguide, which contains intense microwave radiation. When the injection phase is at its optimum, the electrons are able to gain energy from the alternating electromagnetic field. Their final energy depends on the average strength of the field and the length of the waveguide.[43] Such an accelerator is referred to as a *microwave linear accelerator,* or *linac,* and is the most prevalent type of indirect accelerator.[15] It is shown schematically in Figure 4.5.

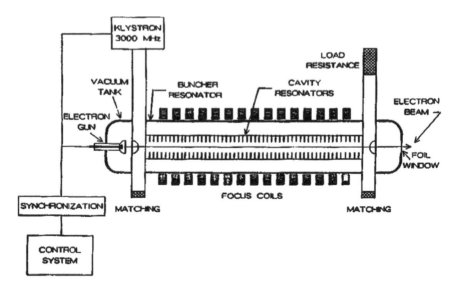

FIGURE 4.5
Schematic of an indirect accelerator (microwave linac). (Adapted from Reference 49.)

Linacs operate with much lower average beam current and power levels than most DC accelerators to attain sufficiently high electron energy. The relatively low voltage of the microwave component and the use of a grounded copper waveguide result in smaller size when compared to DC machines with the same electron energy. The linacs' disadvantage is a low overall electrical efficiency ranging from 20% to 30%.[21]

Low-power linacs are mostly used in cancer therapy and industrial radiography, whereas medium-power linacs are utilized in radiation processing.

4.1.2.1 Traveling Wave Linacs

In traveling wave (TW) linacs, the microwave power is injected in one end of the waveguide and propagates to the other end. At that point, any residual microwave power is dissipated in a resistive load. Low-energy electrons are also injected with the microwave power and travel along in step with the moving electric wave, gaining energy from it continuously.

The peak microwave power must be several MW in order to obtain energy gains of several MeV per meter, and the system must be operated in short, repetitive pulses to keep the average microwave power down to reasonable levels. The electron beam current during the pulse ranges from 0.1 to 1.0 A for an average beam power of 10–20 kW at the electron energy of 10 MeV.[22] Although with optimal beam current, TW linacs can transfer microwave power with an efficiency ranging from 80% to 90%[23] because of the low efficiency of the microwave generation in a klystron, the overall line-to-beam power efficiency is usually below 30%.

4.1.2.2 Standing Wave Linacs

A standing wave (SW) microwave linear accelerator consists of a linear array of resonant cavities that are energized by a common source of microwave power. These cavities are nearly isolated by webs with small-diameter apertures, and the high-energy electron beam passes through these apertures. However, they are coupled through intermediate cavities, which stabilize the microwave phase relationship between the accelerating cavities.

SW linacs have higher electrical impedance or quality factor (Q) than TW linacs with similar energy and beam power ratings.[21] This can provide higher energy gain for the same length of waveguide or a shorter waveguide for the same energy gain, which is an advantage if the accelerator for the given application needs to be compact.

4.1.2.3 Resonant Cavity Accelerators

Resonant cavity accelerators consist of several resonant cavities in series energized by a single S-band klystron using a microwave power distribution system.[24] Another system consists of a single VHF cavity energized by a triode tube that is less expensive than a klystron. The resonant frequency of the latter is about 110 MHz, which is well below the microwave range.[25] Resonant cavity accelerators with electron energies of a few MeV are useful for irradiating thin polymeric products, such as heat-shrinkable tubing and electrical wire.[26]

4.1.2.4 Linear Induction Accelerators

A linear induction accelerator (LIA) accelerates electrons by a series of single-turn, toroidal pulse transformers through which the beam passes. The energy gain in each stage is equal to the voltage applied to the primary winding. The beam acts as a secondary winding. The electrical impedance of a linear induction accelerator is very low. This makes it suitable for accelerating high-peak beam currents. In principle, the electrical efficiency of this accelerator can be substantially greater than that of a microwave linac.[27]

Detailed information on designs, principles of operation, and performance characteristics of electron accelerators is available in several books.[10,11,28,29]

4.1.2.5 Rhodotron Accelerators

A Rhodotron is an electron accelerator based on the principle of recirculation of a beam in successive passes through a single coaxial cavity resonating in the VHF frequency range. This large-diameter cavity operates with a relatively low microwave field, which makes it possible to achieve continuous wave (CW) acceleration of electron beams to high energies.

Rhodotron cavities are shaped as coaxial lines shorted on both ends and resonating in the half-wavelength mode 107.5 or 215 MHz. The beam crosses the diameter of the cavity in the median plane through successive passes (see Figure 4.6).[30] External dipole magnets are used to bend back the electrons emerging from the cavity and redirect them toward the cavity center. A high-power radio frequency (RF) system using a tetrode tube produces the electric field, allowing an energy gain of 0.83–1.17 MeV per crossing. Ten or 12 crossings of the cavity (which means 9 or 11 bending magnets) are therefore required to obtain 10 MeV electron beams at the exit of this accelerator. A very high-power version can produce a 100 mA beam at 5–7 MeV with six passes. With this version, the average electron beam power is in the range of 500–700 kW, which can be used for producing intense X-ray *(bremsstrahlung)* beams. Since the electrons travel along a rose-shaped path (see Figure 4.6), the name Rhodotron was chosen and is derived from the Greek word *rhodos*, which means "rose."[31]

The electron gun is located at the outer wall of the accelerating cavity, and the electrons are injected into the cavity at a voltage of about 35–40 kV. The cavity is cooled by a water jacket on the inner coaxial conductor and at the end flanges and by discrete water channels along the outer diameter. The system is designed to operate with a 2 MW cooling tower up to an outside temperature of 35°C (95°F). Therefore, no water chiller is

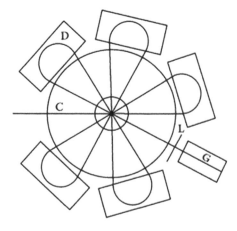

FIGURE 4.6
Schematic of a Rhodotron accelerator. (Courtesy of IBA.)

required.[30] The RF amplifier detects and follows changes in the resonant frequency of the cavity so that accurate control of the cavity temperature is not required.

The electron beam is of very high quality with low-energy spread and low angular divergence.[31] This simplifies the design and operation of external beam transport and scanning systems.[32] The beam delivery system includes a scanning horn with a vacuum-to-atmosphere window using a thin metal foil. An example of a Rhodotron with a scanning horn is in Figure 4.7.

The control system is based on an industrial programmable logic controller (PLC). It includes all software required for the completely automatic operation, maintenance, and troubleshooting of the accelerator.[30]

In addition to providing electron beams with 5 and 10 MeV electron energy outputs for usual EB applications, a Rhodotron can be used to produce *bremsstrahlung X-rays* from a metallic target that can be used for industrial applications.[32,33–35] A Rhodotron is particularly suitable for applications that need a powerful beam (30–200 kW) in the energy range of 1–10 MeV.[36]

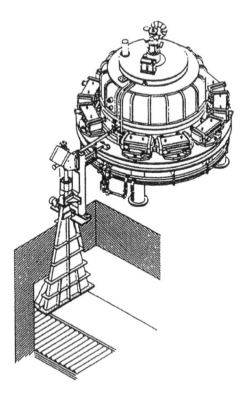

FIGURE 4.7
Example of a Rhodotron system with a scanning horn. (Courtesy of IBA.)

4.1.3 Low-Energy Electron Accelerators

Electron accelerators discussed in previous sections are generating high-energy electrons (10 MeV or even more) and are mainly used to process thick polymeric products, typically up to 20 mm, and for sterilizing medical devices and food processing. However, the vast majority of current industrial irradiation work—such as cross-linking of thin polymeric films and sheets, polymerization, and cross-linking of coatings—is done by low- to moderate-energy accelerators with low electron energies (less than 1.0 MeV).[37] Low-energy accelerators are used, often as reliable computer-controlled subsystems, in coating lines, printing presses, laminating machines, etc. Their operating parameters, such as electron energy, electron beam power, irradiation width, and delivered dose rate, can be matched to the demands of the given industrial process. Such machines are often referred to as *electron processors*. The three most common designs of industrial low-energy electron accelerators are shown in Figure 4.8, and their general operating characteristics are listed in Table 4.1. A comparison of standard and low-voltage EB processors is shown in Table 4.2.

The selection of a suitable electron processor type is dictated by the process parameters. *Scanning*-type electron accelerators use a typical small-diameter electron gun, which generates and shapes a pencil beam. The beam is deflected over the window area by using periodically changing magnetic fields. Carefully tuned deflection produces a very uniform beam current (or dose rate) distribution across the beam exit. Energies up to 1.0 MeV can be reached at medium to high beam current. Due to limited deflection angle,

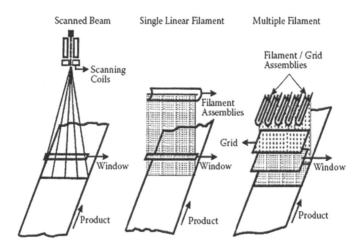

FIGURE 4.8
Common designs of industrial low-energy accelerators. (Adapted from Mehnert, R. et al., *UV & EB Curing Technology & Equipment*, John Wiley & Sons, Chichester/SITA Technology, London, p. 139, 1998.)

TABLE 4.1

General Operating Characteristics of Different Designs of Selected Commercial Low-Energy Industrial EB Processors

Characteristic	Accelerator		
	Single Filament	**Multifilament**	**Single-Stage Scanned Beam**
Accelerating voltage, kV	150–300	150–300	150–300
Working width, mm	150–2000	300–2500	150–2000
Maximum web speed at 10 kGy, m/min	1350	1500	1350
Typical dose variation across the web, %	±10	±8	±4

TABLE 4.2

Comparison of Standard and Low-Voltage EB Processors for a 48-Inch-Wide Web

Dimensions	Standard Processor	Low-Voltage Processor
In-line dimensions, ft (mm)	7 (2135)	4.5 (1370)
Height, ft (mm)	10 (3050)	4.5 (1370)
Depth, ft (mm)	12 (3660)	6 (1830)

Source: Lapin, S.C., *UV+EB Technol.*, 2, 12, 2016.

the length of the scanning horn must be somewhat larger than the length of the exit window.

Single-filament design, consisting of a long linear filament to determine the beam current and beam width, eliminates the need for a beam scanning system. Because the current extractable from a unit length of the single filament is limited, multifilament cathode configurations are often used. This also improves the uniformity of the beam current distribution, which is usually less homogeneous than that obtained in scanning horn processors.[38] The beam width is limited by the necessity to span the filament across the full width of the exit window to a maximum of about 2 m (80 in.). This design is no longer used on currently made machines.

Multifilament design, consisting of an assembly of short, about 20 cm (8 in.) long, filaments set parallel to the web direction, eliminates the problem of filament support and adjustment of the single-filament design. The filaments are mounted between two rigid bars determining the length of the filament assembly. A control and screen grid define the electron optical extraction conditions for each filament into a single gap accelerator. The overlapped beams emitted from adjacent cathode filaments form a uniform electron cloud before the acceleration takes place. This design is capable of generating high beam power, dose uniformity, and a wider beam than the single-filament design.[38] A multifilament array of an accelerator is shown in Figure 4.9.

FIGURE 4.9
Multifilament array of an EB accelerator. (Courtesy of M. R. Cleland.)

4.1.3.1 Single-Stage Scanned Beam Accelerator

The accelerator for the electron beam processor produced by Electron Crosslinking AB in Halmstadt, Sweden, originally developed by Polymer Physik in Tübingen, Germany, uses a classical triode system as electron gun, a single electrode gap, and a beam focusing and deflection system. The acceleration voltage depends on the application and ranges from 150 to 300 kV.[39] The processor is shown in Figure 4.10.

The electron gun consists of a spiral-shaped tungsten cathode and a Wehnelt cylinder. These two components not only constitute the electrodes of the acceleration gap but also form the optical assembly to control and shape

FIGURE 4.10
Single-stage scanned beam accelerator with a range of accelerating voltages from 80 to 300 kV. (Courtesy of Elektron Crosslinking AB.)

the electron beam. Current signals are linear and have a repetition frequency of about 800 Hz. They are used to deflect the electron beam horizontally and vertically over the exit window plane. The scanner may be equipped by two cathodes for maximum output. Then, the width of the exit window is more than double that of a standard unit with a single cathode. The exit window containing the 12–15 μm thick titanium foil is relatively large to ensure an effective cooling of the foil.

Nissin High Voltage Co. of Japan also offers machines with electron scanners. The acceleration voltages in these machines are in the 300–500 kV range with beam powers up to 65 kW. The maximum beam width is 1,200 mm (48 in.).[67] An example of a machine of this is in Figure 4.11.

4.1.3.2 Linear Cathode Electron Accelerators

A linear cathode accelerator employs a cylindrical vacuum chamber in which a longitudinal heated tungsten filament cathode is raised to the negative accelerating potential. The electrons emitted from the cathode are accelerated in a single step to the exit window. The exit window as a part of the vacuum chamber is kept at ground potential. The accelerated electrons penetrate the thin titanium window and enter the process zone. A schematic of a linear cathode accelerator is in Figure 4.12. A linear cathode processor using a single filament has a dose-speed capacity of about 450 m/min (1,500 ft/min) at 10 kGy. In the actual commercial equipment, the cathode assembly contains up to four filaments, which are capable of providing a dose-speed capacity

FIGURE 4.11
Self-shielding EB processor with the accelerating voltages of 800 kV. (Courtesy of Nissin Electric.)

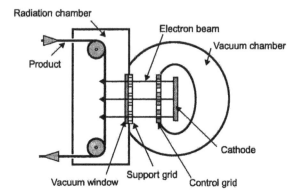

FIGURE 4.12
Schematic of a linear cathode accelerator. (Courtesy of Elektron Crosslinking AB.)

of up to 1,350 m/min (4,500 ft/min) at 10 kGy or a beam current of several hundred mA.[40] The processor is supplied with a protective shielding against X-rays in the form of a lead slab. The X-ray shielding and product handling system are designed to match the specific process demands.

The process control is provided by programmable process controllers (PLCs), which are flexible and upgradable. The addition of man-machine interfaces (MMIs) based on personal computers and common software systems allows not only system control and interlocks, but also a history of operation, communication, and archiving.[41] A series of processors of this design were manufactured by Energy Sciences, Inc. in Wilmington, Massachusetts, under the brand Electrocurtain® in the 1970s and 1980s.

Another low-energy processor can be described as a vacuum diode operating in the current saturation mode, i.e., the cathode is a directly heated tungsten filament without a control grid. The electrons emitted from the cathode are shaped into electron beam bundles by a forming electrode and accelerated to the exit window. This diode configuration ensures that practically all electrons emitted are accelerated. Cathode heating can be accomplished through the high-voltage cable. This allows moderate electron currents at low heating power.[42] An example of such a processor was EBOGEN, produced by igm Robotersysteme in Munich, Germany (Figures 4.13 and 4.14). At this writing, the equipment has been reportedly modified and is produced by Steigerwald Strahltechnik in Maisach, Germany, under the trade name EBOCAM®. It is used mainly for welding and drilling metals.

Low-energy self-shielded electron beam machines have been used in industrial applications for more than 30 years. The first-generation system operated mostly within the accelerating voltage range from about 175 to 300 kV processing webs up to 3,000 mm (120 in.) wide. Typical applications

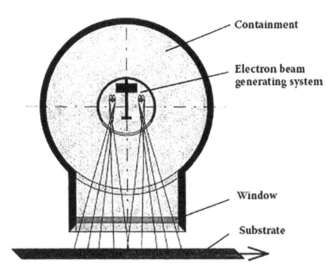

FIGURE 4.13
Schematic of the processor EBOGEN. (Courtesy of igm Robotersysteme.)

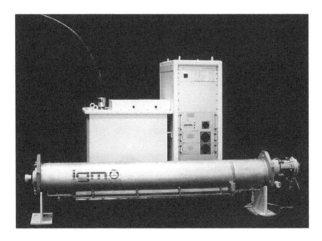

FIGURE 4.14
EBOGEN processing unit. (Courtesy of igm Robotersysteme.)

included cross-linking thermoplastic films for heat-shrinking packaging and tubing as well as processing plies for tires. In some cases they were also used for curing coatings and inks for printing. In the latter applications such large machines were only useful for long single material runs. A new generation of smaller EB systems operating in the range from 80 to 125 kV was introduced in the early 1990s. This automatically meant decreasing size and weight and accelerating voltage used.

Lower accelerating voltage means a less expensive power supply, smaller size accelerator head, and less X-ray shielding and reduced amount of nitrogen for inerting. For example, a small 75 kV processor is capable of curing coatings up to 0.001 in. (25 µm) thick (at density 1). Complete cure is achieved at doses 10–25 kGy, which is considerably lower than the typical 30 kGy required for the 150 kV equipment. Compared to a standard processor, the total volume is 50% smaller. Smaller machines with accelerating voltages in the range from 70 to 125 kV can be used in coil coating, web offset presses, flexo presses, laminating presses, and specialty cross-linking.

The penetration depth of electrons into the substrate decreases at lower energy. Also, the surface dose per mA of beam current increases, and thin coatings and inks can be cured with higher efficiency and speed. The higher surface dose produces a higher concentration of radicals, which in turn reduces the oxygen inhibition.

4.1.3.3 Multifilament Linear Cathode Electron Accelerators

In this design, a multiple emitter assembly with heated tungsten filaments placed in parallel to the product's direction is used. The beam current is controlled by molybdenum grids held at a common potential. A planar screen grid is placed below the control grids. In the field-free region between the grids, a highly uniform electron density is generated prior to acceleration. The acceleration of electrons occurs between the planar screen grid and the grounded window. This design is represented by the BroadBeam™ processor that was manufactured by the now-defunct RPC Industries in Hayward, California. Currently, this type of equipment with additional design improvements is produced by PCT Ebeam and Integration, LLC in Davenport, Iowa (Figure 4.15). The machine shown is the *Invictus Series P2012094* with

FIGURE 4.15
PCT Invictus series system P22012044. (Courtesy of PCT Ebeam and Integration, LLC.)

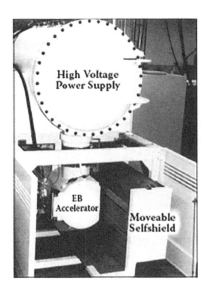

FIGURE 4.16
Low-energy self-shielding electron beam processor. (Courtesy of Energy Sciences, Inc.)

maximum accelerating voltage of 300 kV, maximum web width of 1500 mm (60 in.), and an Integrated Shield Roll.

A new window design (cooling and support structure) combined with a multifilament cathode arrangement led to the development of low-energy, low-cost electron processors, such as Electrocure™ and EZCure™ manufactured by Energy Sciences, Inc. in Wilmington, Massachusetts (Figure 4.16). For example, a 150 kV processor is capable of a dose-speed capacity of 1,200 m/min (3,950 ft/min) at 10 kGy. An even more powerful 125 kV, 200 kW processor is capable of curing coatings on a web 1,650 mm (66 in.) wide with a dose-speed capacity of 2,250 m/min (7,400 ft/min) at 10 kGy (Chrusciel, J., private communication).[43] Smaller, lower-voltage processors operating between 90 and 110 kV accelerating voltages are capable of speeds up to 400 m/min (1,310 ft/min) (Chrusciel, J., private communication). Another view of an EZCure™ machine is in Figure 4.17. The manufacturer did not provide details on dimensions and performance for this machine, so those who are interested are advised to contact the company at www.ebeam.com.

Additional currently available smaller commercial EB processors are:

- *Core 100* is the model of an EB equipment (see also Figure 4.16) with maximum accelerating voltage of 110 kV, maximum web width of 760 mm (30 in.), and an Integrated Shield Roll produced by PCT Ebeam and Integration, LLC), shown in Figure 4.18.

FIGURE 4.17
EZCure FC system. (Courtesy of Energy Sciences, Inc.)

FIGURE 4.18
Core 100 EB system. (Courtesy of PCT Ebeam and Integration, LLC.)

- Patented *In-Line System* by Swedish company Crosslinking AB suitable for printing, curing, and lamination shown in Figure 4.19. This machine has maximum acceleration voltage up to 250 kV, maximum speed 300 m/ min (984 ft/min), and can work with webs up to 680 mm (27 in.) wide.

FIGURE 4.19
In-line system. (Courtesy of Crosslinking AB.)

- *EC-Compact* processor for printing, curing, and sterilization made and offered by Crosslinking AB with acceleration voltage in the range of 70–150 kV with working web width of 100–300 mm (4–12 in.), web speed at 10 kGy 900 m/min (2950 ft/min), and dose distribution less than ±7.5%. This processor is shown in Figure 4.20.

FIGURE 4.20
EC-compact processor. (Courtesy of Crosslinking AB.)

4.1.4 Sealed Tube Lamp Accelerators

In the effort to simplify electron beam equipment and make it competitive with UV and UV-EB lamps and small second-generation UV reactors, A. Avnery developed and patented an EB emitter, which used sealed window foil and maintained a permanent vacuum, eliminating the need for vacuum pumps. This emitter is essentially a sealed vacuum tube with a 2.5 μm thick silica ceramic window or a 6 μm thick titanium window as the beam exit. The emitters operate at a voltage range of 50–150 kV and can be connected into modules to cure wide webs.[44] Rather than replacing foils, new emitters were designed to be plugged in in the event of the emitter failing. Failed emitters were returned and rebuilt in the factory. This concept was essentially a lamp very different from a traditional EB accelerator.[45] Initially, these emitters were available in widths of 250 mm (10 in.) and 400 mm (16 in.). A schematic of the patented emitter is in Figure 4.21a, and the actual 10 in. diameter emitter is in Figure 4.21b. The 400 mm emitter is in Figure 4.22. A review of this reactor was prepared by Berejka.[46] The inventor started a company named AEB. The company produced and sold several hundred emitters but eventually went out of business.

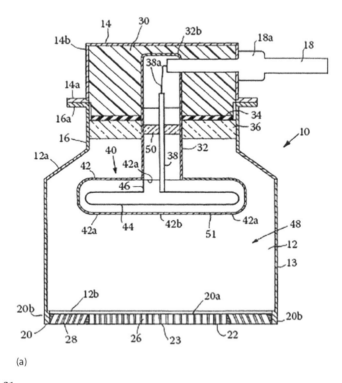

(a)

FIGURE 4.21
(a) Schematic of a sealed low-voltage electron beam emitter. (US Patent 6,545,398)

(Continued)

FIGURE 4.21 (Continued)
(b) Actual sealed AEB low-voltage 250 mm (10 in.) emitter. (Courtesy of AEB Inc.)

FIGURE 4.22
Sealed AEB low-voltage 400 mm (16 in.) emitter. (Courtesy AEB Inc.)

As of this writing, METALL + PLASTIC company of Germany owns the rights to manufacture all of the AEB technology and has continued to manufacture as well as further develop the 250 and 400 mm emitters (www.metall-plastic.com/en) for curing and sterilization. Likewise, the Japanese company Hitachi Zosen Corporation owns the rights to manufacture all of the AEB technology and currently offers only an electron beam sterilization system (www.hitachizosen.co.jp/english/products/products041.html).

After the demise of the AEB company in the USA, Swiss company Comet developed its own industrial sealed tube lamp. The sealed tube has a tri-ode design with one grid at negative voltage based on the 450 kV bipolar X-ray tube design, produced by that company.[47] These lamps are available in 270 mm (11 in.) and 400 mm (16 in.) width with accelerating voltages ranging from 80 to 200 kV and used for curing and sterilization. The Comet *ebeam* system (EB lamp, power supply and cooling unit) is in Figure 4.23. Another similar sealed tube EB equipment, *EB-ENGINE*, is being offered by Japanese company Hamamatsu, as shown in Figure 4.24. It is reported to be under development and working with 50–100 kV accelerating voltage and having maximum processing capability of 3,000 kGy m/min (30 kGy @ 100 m/min).

The possible applications of these EB systems are[45]:

- EB Laboratory systems
- In-line package sterilization
- Narrow web printing
- EB curing on three-dimensional parts

The compact lamps can be configured and used together with automated product handling to present all surfaces that require curing within the needed proximity of the beam window.

FIGURE 4.23
Comet ebeam system. (Courtesy of Comet AG.)

FIGURE 4.24
Hamamatsu EB-ENGINE. (Courtesy of Hamamatsu Company.)

Several companies have developed and are offering EB equipment for laboratory work intended mainly for material and process research.[48] Few of the companies are listed below:

- Comet AG e-beam (Test Lab)
- Elektron Crosslinking AB (EC Lab 400)
- Energy Sciences, Inc. (EZLab)

References

1. Beying, A., *RadTech Europe '97, Conference Proceedings* (June 16–18, Lyon, France), p. 77 (1997).
2. Lauppi, U. V., *RadTech Europe '97, Conference Proceedings* (June 16–18, Lyon, France), p. 96 (1997).
3. Mehnert, R., Pincus, A., Janorsky, I., Stowe, R., and Berejka, A., *UV & EB Curing Technology & Equipment*, John Wiley & Sons, Chichester/SITA Technology, London, p. 141 (1998).
4. Mehnert, R., Pincus, A., Janorsky, I., Stowe, R., and Berejka, A., *UV & EB Curing Technology & Equipment*, John Wiley & Sons, Chichester/SITA Technology, London, p. 143 (1998).
5. Bradley, R., *Radiation Technology Handbook*, Marcel Dekker, New York, 1984, p. 37 (1983).
6. Artandi, C., and Van Winkle Jr., W., *Nucleonics*, 17, p. 86 (1959).
7. Charlton, E. E., Westendorp, W. F., and Dempster, L. E., *J. Appl. Phys.*, 10, p. 374 (1939).

8. Cleland, M. R., in *Radiation Processing of Polymers* (Singh, A., and Silverman, J., Eds.), Carl Hanser Verlag, Munich, Chapter 3, p. 28 (1992).
9. Sakamoto, I., Mizusawa, K., *Radiat. Phys. Chem.*, 18, p. 1341 (1981).
10. Scharf, W., *Particle Accelerators and Their Uses*, Harwood Academic Publishers, New York (1986).
11. Abramyan, E. A., *Industrial Electron Accelerators and Applications*, Hemisphere, Washington, DC (1988).
12. Van de Graaff, R. J., U.S. Patents 3,187,208 (1965), 3,289,066 (1966), 3,323,069 (1967).
13. Emanuelson, R., Fernald, R., and Schmidt, C., *Radiat. Phys. Chem*, 14, p. 343 (1979).
14. Sakamoto, I., et al., *Radiat. Phys. Chem.*, 25, p. 911 (1985).
15. Cleland, M. R., in *Radiation Processing of Polymers* (Singh, A., and Silverman, J., Eds.), Carl Hanser Verlag, Munich, Chapter 3, p. 29 (1992).
16. Cleland, M. R., Thompson, C. C., and Malone, H. F., *Radiat. Phys. Chem.*, 9, p. 547 (1977).
17. Thompson, C. C., and Cleland, M. R., *Nucl. Instrum. Meth.*, B40/41, p. 1137 (1989).
18. Morganstern, K. H., in *Proceedings of the Conference on Sterilization of Medical Products*, Vol. 4, Johnson & Johnson, Polyscience Publications, Montreal (1986).
19. Odera, M., Nagakura, K., and Tanaka, Y., *Radiat. Phys. Chem.*, 35, p. 534 (1990).
20. Cleland, M. R., Thompson, C. C., Strelczyk, M., and Sloan, D. P., *Radiat. Phys. Chem.*, 35, p. 632 (1990).
21. Cleland, M. R., in *Radiation Processing of Polymers* (Singh, A., and Silverman, J., Eds.), Carl Hanser Verlag, Munich, Chapter 3, p. 30 (1992).
22. Haimson, J., *Proceedings of the Conference on Sterilization by Ionizing Radiation 1*, Johnson & Johnson, Multiscience Publications Ltd., Montreal (1974).
23. Haimson, J., *IEEE Trans. Nucl. Sci*, NS-22, p. 1303 (1975).
24. Anonymous, *Beta-Gamma*, 3, p. 38 (1989).
25. Auslender, V. L., et al., U.S. Patent 4,140,942 (1979).
26. Cleland, M. R., in *Radiation Processing of Polymers* (Singh, A., and Silverman, J., Eds.), Carl Hanser Verlag, Munich, Chapter 3, p. 34 (1992).
27. Barletta, W. A., *Beam Research Program, Energy and Technology Review*, Lawrence Livermore National Laboratory, Livermore, CA, LLL TB 63 (1984).
28. Humphries Jr., S., *Principles of Charged Particles Acceleration*, John Wiley & Sons, New York (1986).
29. Lapostolle, P. M., and Septier, A. L., *Linear Accelerators*, North Holland Publishers, Amsterdam (1970).
30. Abs, M., Jongen, Y., Poncelet, E., and Bol, J.-L., *Radiat. Phys. Chem.*, 71, p. 285 (2004).
31. Lancker, M. V., Herer, A., Cleland, M. R., Jongen, Y., and Abs, M., *Nucl. Instrum. Meth.*, B151, p. 242 (1999).
32. Jongen, Y., Abs, M., Genin, F., Nguyen, A., Capdevilla, J. M., and Defrise, D., *Nucl. Instrum. Meth.*, B79, p. 8650 (1993).
33. Korenev, S., *Radiat. Phys. Chem.*, 71, p. 535 (2004).
34. Korenev, S., *Radiat. Phys. Chem.*, 71, p. 277 (2009).
35. Meissner, J., Abs, M., Cleland, M. R., Herer, A. S., Jongen, Y., Kuntz, F., and Strasser, A., *Radiat. Phys. Chem.*, 57, p. 647 (2000).
36. Bassaler, J. M., Capdevilla, J. M., Gal, D., Lainé, F., Nguyen, A., Nicolai, J. P., and Uniastowski, K., *Nucl. Instrum. Meth.*, B68, Issue 1–4, p. 92 (1992).

37. Singh, A., and Silverman, J., in *Radiation Processing of Polymers* (Singh, A., and Silverman, J., Eds.), Carl Hanser Verlag, Munich, Chapter 1, p. 10 (1992).
38. Mehnert, R., Pincus, A., Janorsky, I., Stowe, R., and Berejka, A., *UV & EB Curing Technology & Equipment*, John Wiley & Sons, Chichester/SITA Technology, London, p. 139 (1998).
39. Holl, P., *Radiation Physics and Chemistry*, 25, p. 665 (1985).
40. Mehnert, R., Pincus, A., Janorsky, I., Stowe, R., and Berejka, A., *UV & EB Curing Technology & Equipment*, John Wiley & Sons, Chichester/SITA Technology, London, p. 150 (1998).
41. Meskan, D. A., and Klein, F. A., *Proceedings, RadTech Europe '97* (June 16–18, Lyon France), p. 114 (1997).
42. Schwab, U., *Proceedings RadTech Europe '97* (June 16–18, Lyon, France), p. 114 (1997).
43. Mehnert, R., Pincus, A., Janorsky, I., Stowe, R., and Berejka, A., *UV & EB Curing Technology & Equipment*, John Wiley & Sons, Chichester/SITA Technology, London, p. 155 (1998).
44. Drobny, J. G., *Ionizing Radiation and Polymers: Principles, Technology and Applications*, Elsevier, Oxford, UK, p. 75 (2013).
45. Lapin, S. C., "What types of applications are enabled by sealed tube EB lamp technology?" *UV+EB Technology*, 2, No. 2, p. 12 (2016).
46. Berejka, A.J., *Radiat. Phys. Chem.*, 71, p.309 (2003)
47. Haag, W., "Sealed electron beam emitter for use in narrow web curing, sterilization and laboratory application," in *Proceedings from RadTech UV & EB Conference* (April 30–May 2, Chicago, IL) (2012).
48. Drobny, J. G. *Ionizing Radiation and Polymers: Principles, Technology and Applications*, Chapter 3, Elsevier, Oxford, UK (2013).

5

UV Radiation Processes (with Ruben Rivera)

Electromagnetic radiation (or light) can be used as the energy source for a variety of processes involving functional monomers, oligomers, and polymers. In these processes, it is utilized mainly to effect the formation of new chemical bonds. The light used in polymeric systems typically has wavelengths ranging in the ultraviolet spectral range extending from 200 to 400 nm, although in some special cases it may include wavelength in the visible range up to 760 nm. Xenon lamps provide significant emissions in the range from 450 to 550 nm.[1] Because the energy of radiation accelerates with increasing frequency (or decreasing wavelength), radiation with short waves contains a large amount of energy. This energy is capable of bringing about certain chemical reactions in a system that is sensitive to light, which means that it has the ability to absorb it. Then the absorbed energy can generate species, which are capable of initiating polymerization or cross-linking reactions. For the most part, UV radiation is used to convert reactive liquid monomers and oligomers to solids. This process involves essentially polymerization and sometimes polymerization and cross-linking simultaneously. Monomers and oligomers with a functionality of two form linear polymers, whereas multifunctional ones (with a functionality of more than two) yield three-dimensional cross-linked networks. Cross-linking imparts solvent resistance, increases hardness, and improves heat resistance. The main practical utility of these reactions is in curing coatings and inks and in photo-imaging. Cross-linking of already existing polymers is also a viable process and is currently a very active field of research.[1]

5.1 Basic Concepts

The processes of photochemistry are the same for polymers and small molecules.[2–6] The *Grotthus-Draper law* states that no photochemical reactions can occur unless a photon of light is absorbed. This means, for example, that many commercial plastics transparent in the near UV can undergo photodegradation only as a result of the absorption of light by impurities. The intensity of any light absorbed by a light-absorbing species (chromophores) follows the Beer-Lambert law:

$$I = I_0 10^{-\varepsilon cd} \tag{5.1}$$

where I_0 is the intensity of the incident light, I is the intensity of transmitted light, ε is the molar extinction coefficient (cm^1 mol^1), c is the concentration of absorbing species, and d is the optical path length. *Absorbance* (A), or optical density, is defined as $-\log (I/I_0)$; then $A = \varepsilon cd$. Typical chromophoric groups for UV light are –C=O, –ROOH, and aromatic groups. These extend the absorption of monomers, oligomers, and polymers into the UV light range.[7] Figure 5.1 shows an energy level (Jablonski) diagram for a ketone, a common chromophore in polymers.[8] Several intramolecular processes are shown competing with photochemical reactions, including reemission of a photon as fluorescence or phosphorescence, radiationless decay to the ground state, and crossing from one excited state to another.

The ground states of almost all organic compounds have all electron spins paired. Absorption of a photon promotes an electron from the singlet state, S_0, to a higher-energy singlet state, $S_1, S_2,...,S_n$, numbered in the order of increasing energy above the ground state. A change in the spin state of an electronically excited molecule is called *intersystem crossing;* it produces triplet species, $T_1, T_2,...,T_n$, with two unpaired spins.[9] A triplet state is always lower in energy than the corresponding singlet state. Singlet and triplet states may emit light and return to the ground state (Andrzejewska, E., private communication).

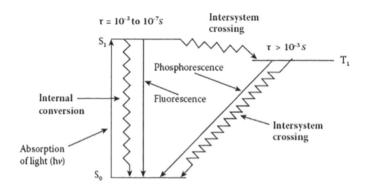

FIGURE 5.1
Primary processes occurring in the excited state of a UV radical photoinitiator. (Adapted from Jablonski, A.Z., *Physik*, 94, 38, 1935.)

To put it simply:

- The absorption of a photon by a chromophore brings about a transition into the excited singlet state.
- Generally, the excited molecule has two possibilities to emit the absorbed energy: it can either return to the ground state by emitting energy by fluorescence or a nonradiative transition, or it can cross over to the excited triplet state (Andrzejewska, E., private communication).
- Molecules in the triplet state are biradicals, which can, if the energy is high enough for breaking a bond, form free radicals. The free radicals can then initiate the polymerization or cross-linking reaction.

The main decay processes to the ground state shown in Figure 5.1, which is essentially an energy diagram for the different electronic states, are:

- Radiative processes:
- Absorption: $S_0 + hv \rightarrow S_1$
- Fluorescence: $S_1 + S_0 + hv'$
- Phosphorescence: $T_1 \rightarrow S_0 + hv''$

where h is the Planck's constant, and v, v', and v'' are the respective frequencies of the absorbed or emitted light.

- Radiationless processes:
- Internal conversion: $S_1 \rightarrow S_0 + heat$
- Intersystem crossing: $T_1 \rightarrow S_0$ or $S_1 \rightarrow T_1$

The result of a photochemical reaction involving monomers, oligomers, and polymers depends on the chemical nature of the material, wavelength of the light, and other components of the system. Ultraviolet, visible, and laser light can polymerize functional monomers, cross-link polymers,[1] or degrade them, particularly in the presence of oxygen.[10]

5.2 Photoinitiators and Photosensitizers

There are essentially two types of compounds that are used in the UV curing process to absorb the light and generate reactive species. These are photoinitiators and photosensitizers.

Photoinitiator (PI) is a compound-generating reactive species that will initiate polymerization or cross-linking. *Photosensitizer* (S) is a compound that will energize certain species that will in turn lead to production of reactive species. It is a molecule that absorbs light usually at longer wavelengths and transfers energy to a photoinitiator to generate free radicals or ions.

$$PI \rightarrow PI^* \rightarrow \text{Reactive species (free radicals or ions), or } S \rightarrow S^*$$
$$S^* + PI \rightarrow S + PI^* \qquad \text{Energy transfer to photoinitiator}$$

Thus, photosensitizers are useful mainly by being capable of extending the spectral sensitivity of certain photoinitiators under specific conditions.

The function of a photoinitiator is:

- Absorbing the incident UV radiation
- Generating reactive species (free radicals or ions)
- Initiating photopolymerization

In the UV curing process, photons from the UV source are absorbed by a chromophoric site of a molecule in a single event. The chromophore is a part of the photoinitiator. The light absorption by the photoinitiator requires that an emission light from the light source overlap with an absorption band of the photoinitiator.

The photon absorption follows the Beer-Lambert law (see above). The number of photons I present at depth l from the surface is given as a function of the optical absorbance, A, normalized to the initial number of photons, I_0:

$$\log(I_0/I) = A = \varepsilon \, [PI] \; l \qquad (5.2)$$

where [PI] is the concentration of the photoinitiator. The quantity l is also termed the *photon penetration path*. In general, upon exposure to UV radiant energy, a photoinitiator can generate free radicals or ions, as pointed out earlier. These are generated at a rapid rate, and their depth profile corresponds to the inverse photon penetration profile. Similar to electron penetration, the final cure profile often deviates from the initial radical or ion distribution, since they can live much longer than the exposure time. The mechanisms of the processes for the generation of reactive species are discussed in detail in Davidson.[11]

Depending on the type of reactive species generated upon exposure to UV light, photoinitiators are classified as free radical, cationic, and anionic.

5.2.1 Free Radical Photoinitiators

The UV curing of certain monomers, such as acrylate, methacrylate, and maleate/vinyl ether systems, is initiated by free radicals. In all practical cases, the initiating radicals are generated from electronically excited photoinitiator molecules[12,13]. A photoinitiator molecule is excited into the singlet state by the absorption of a photon. The formation of a radical occurs via a triplet state. Radical formation occurs via two possible reaction sequences. They are designated as Norrish Type I and Type II reactions. In a Type I reaction, the photoinitiator triplet state decays into a radical pair by homolytic decomposition and directly forms radicals capable of initiating polymerization. The absorbed radiation causes bond breakage to take place between a carbonyl group and an adjacent carbon. In a Type II Norrish reaction, triplet states of ketones possessing an α-hydrogen preferably react with suitable hydrogen-donating compounds by hydrogen abstraction. The resulting radical pair can be generated either by a homolytic cleavage of the R-H bond or via an intermediate charge transfer complex followed by proton transfer.[14] The lifetime of the excited initiator species is very short, generally less than 10^{-6} s. During this time, it may be partitioned essentially between two processes: it can decay back to the original state with emission of light or heat, or it can yield a reactive intermediate (free radical or ion), which in turn can react with another free radical or initiate polymerization of a monomer.[15]

Free radical photoinitiators are classified by their chemical nature as Type I and Type II; however, there are a few systems with different chemistry, for example, borate salt initiators that depend on intermolecular/intramolecular electron transfer,[16] that do not fit into either category.

5.2.1.1 Type I Photoinitiators

Type I initiators are compounds that upon irradiation undergo a cleavage reaction (α- or β-cleavage) to generate two radicals[17]:

The chromophore in this type of photoinitiator is frequently an aromatic carbonyl. The benzoyl radical is the major initiating species, while the other fragment may also contribute to the initiation, in some cases. The most efficient Type I initiators are benzoin ether derivatives, benzil ketals, hydroxylalkylphenones, α-aminoketones, and acylphosphine oxides.[18] Substituents on the aromatic carbonyl influence the absorption.

5.2.1.2 Type II Photoinitiators

A large number of aromatic ketones (benzophenone, substituted benzophenone, benzil, fluorenone, camphorquinone, xanthone, and thioxantone) (Andrzejewska, E., private communication) will act as Type II initiators, with their performance being dramatically enhanced by the use of tertiary amine synergists. There are essentially two classes of amine synergists, aliphatic and aromatic, with different characteristics and uses.[19] Aliphatic amines are transparent down to about 260 nm, and consequently, a coating containing such amine can be cured with UV light down to that wavelength. Aromatic amine synergists display strong absorption around 300 nm and consequently screen much of the UV light.[20] Therefore, in cures with benzophenone it is best to use an aliphatic amine, whereas thioxanthones, because of their strong absorption at wavelengths greater than 340 nm, can be used successfully with an aromatic amine.[19]

Combinations of Type I and Type II photoinitiators are used to reduce air inhibition in the absence of amines.[21] A possible explanation of the synergistic effect is that the photo-cleavage of the Type I initiator yields free radicals that react with oxygen to generate peroxy-radicals (R–O–O•). These radicals are poor initiators for usual radical polymerization, but they abstract hydrogen from monomers and oligomers, which regenerates radicals (chain process). Oxygen depletion enhances the effectiveness of Type II initiators as an energy transfer agent for decomposing hydroperoxides into alkoxy (RO•) and hydroxy (HO•) radicals.[22] In contrast to peroxy radicals, they are highly effective initiators for acrylate polymerization. In the overall process, oxygen is used to produce initiating radicals.

Some examples of Type I and Type II photoinitiators are shown in Tables 5.1 and 5.2, respectively.

5.2.2 Free Radical Photoinitiators for UV-LED Curing

The curing technology using ultraviolet light-emitting diodes has been growing rapidly during the last several years. As pointed out in Section 2.1.4, UV light-emitted diodes produce the UV energy in a very different way than traditional mercury arc lamps, and the UV light produced by them exhibits very different properties. The emitted wavelengths of UV-LEDs are represented by narrow bell-shaped distributions with a nominal irradiance peak. The exact wavelength where the peak occurs and the shape of the distribution are entirely dependent on the structure of the material used—as mentioned earlier—and it is not adjustable, as the diodes are fabricated. While arc and microwave UV systems are broadband emitters with a range of output between 200 and 445 nm, common wavelength peaks for UV-LED systems are 365, 385, 395, and 405 nm. The consequence of the difference between the properties of the two types of lamps (peak intensity and peak wavelength) is that often the UV-LED requires different photoinitiators, as pointed out in Section 3.1.7.

Most photoinitiators have a primary absorbance range below the 365/395 nm peak LED lamp wavelengths. However, UV/LED lamps do not have a purely monochromatic spectrum, and most photoinitiators have broad absorption bands, which are often overlooked when only the maxima are considered. Several photoinitiators do absorb in areas of the spectrum around and above 365 and/or 395 nm. Typically, a 395 nm UV/LED lamp produces 10X the power/area (peak irradiance)

TABLE 5.1

Examples of Type I Photoinitiators

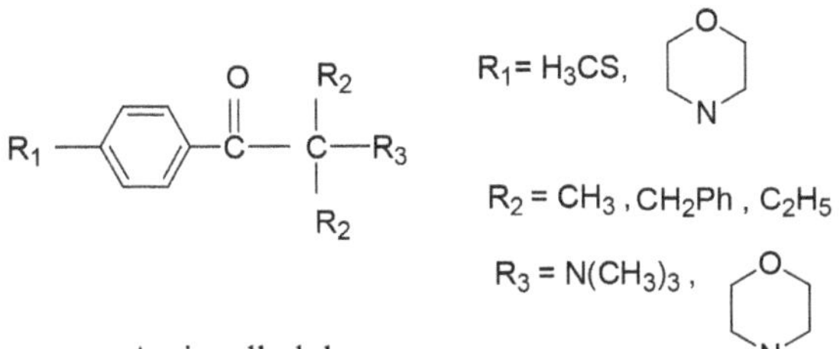

$R_1 = H_3CS$,

Aminoalkylphenones

$R_2 = CH_3 , CH_2Ph , C_2H_5$

$R_3 = N(CH_3)_3$,

Benzoin ethers

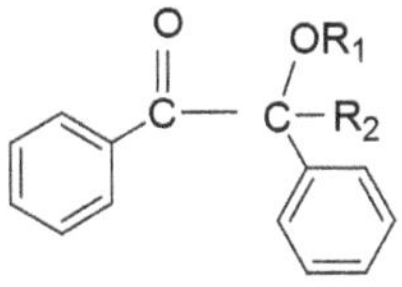

$R_1 = H$ or alkyl

$R_2 = H$ or substituted alkyl

Acetophenones

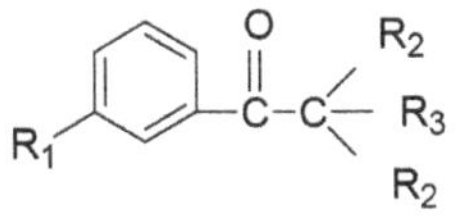

$R_1 = H, iso\text{-}C_3H_7, HOCH_2CH_2O$

$R_2 = CH_3 , OCH_3 , OC_2H_5$

$R_3 = H, Ph, OH$

Acylphosphines

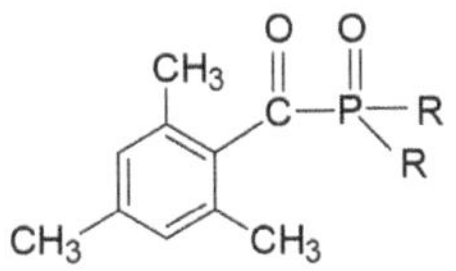

$R = C_6H_5 , OCH_3$

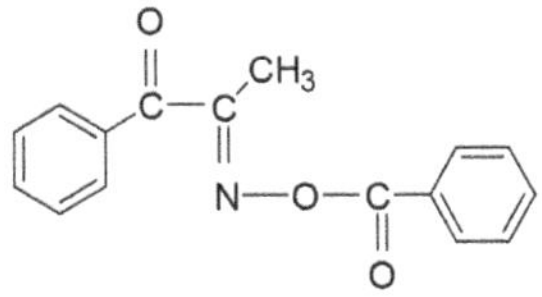

Benzoyloximes

than that produced by a 365 nm lamp. Also, note that the 395 nm lamp provides much greater peak irradiance than the traditional Hg lamp as well. Considering the band overlap potential, it should be possible to find one or more photoinitiators (or, more likely, a combination of photoinitiators) that can provide sufficient free radical flux to initiate an efficient

TABLE 5.2

Examples of Type II Photoinitiators

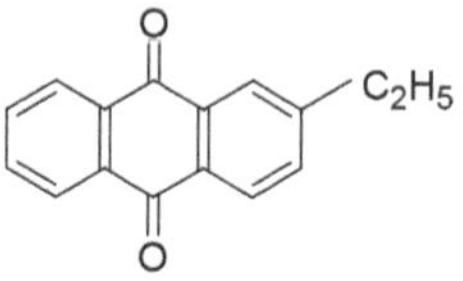

Anthrachinones

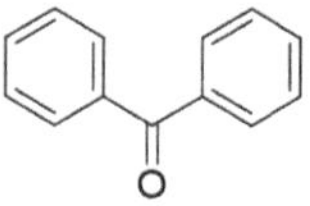

Benzophenone

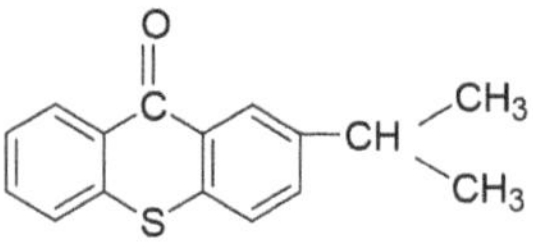

2-Isopropyl thioxanthone (ITX)

polymerization, even in pigmented systems.[23] In fact, some photoinitiator manufacturers offer products that are blends of specific photoinitiators. From numerous reports it appears that currently the UV-LED curing processes are using predominantly free radical chemistry, which means that the photoinitiators suitable for them are Type I and Type II, like in the processes based on mercury arc lamps. As mentioned before, these must be matched with the given peak wavelength being used. The increasing use of UV-EB technology in industrial inks and coating requires a fair amount of research, and there are many reports providing results that are sometimes confusing. For that reason, we are presenting only a simplified discussion including only several photoinitiators and general methods for UV-LED curing. Some authors claim that the "best" photoinitiator depends on the application. In some cases, it is a drop in replacement to conventional UV lamps; in other cases, more work is required.[24] The readers are advised to study the sources in listed references and contact manufacturers of photoinitiators.

The Type I initiators for UV-LED cure exhibit α (unimolecular) cleavage and few examples are amino alkyl phenones, benzoin ethers, and acyl phosphines (see Table 5.3). Some of them exhibit photo-bleaching.[25]

The Type II initiators for UV-LED cure exhibit intermolecular cleavage and require hydrogen donors, such as thiols, amines, and ethers. They produce improved surface cure by mitigation of oxygen inhibition. Examples are benzophenones, thioxanthones, and phenyl glyoxylates (see Table 5.4).[25]

Details on the photoinitiators for UV-LED cure are in References.[23–30]

TABLE 5.3

Examples of Type I Photoinitiators for UV-LED Cure

Abbreviation	Description	Maximum Wavelength, nm	Cutoff Wavelength, ~nm
TPO	2,4,6-trimethylbenzoyldiphenylphosphine oxide	298	420
TPO-L	Ethyl(2,4,6-trimethylbenzoyl) phenyl phosphinate	290	430
BAPO	Phenyl bis(2,4,6-trimethylbenzoyl)-phosphine oxide	281	440
BDMP	2-benzyl-2-dimethylamino-4-morpholinobutyrophenone	306	390

Source: Arcenaux, J.A. et al., UV LED Cure Applications, *Big Ideas Conference*, Redondo Beach, CA, 2019.

TABLE 5.4

Examples of Type II Photoinitiators for UV-LED Cure

Abbreviation	Description	Maximum Wavelength, nm	Cutoff Wavelength, ~nm
BP	Benzophenone	252	320
BMS	4-benzoyl-4'-methyldiphenyl sulfide	315	400
MBF	Methylbenzoylformate	244	320
ITX	2-isopropylthioxanthone	383	420
DETX	2,4-diethylthioxanthone	386	420

Source: Arcenaux, J.A. et al., UV LED Cure Applications, *Big Ideas Conference*, Redondo Beach, CA, 2019.

5.2.3 Cationic Photoinitiators

Cationic photoinitiators are compounds that, under the influence of UV or visible radiation, release an acid, which in turn catalyzes the desired polymerization process.[31] Initially, diazonium salts were used, but they were replaced by more thermally stable iodonium and sulfonium.[32] Examples of cationic photoinitiators are in Table 5.5.

5.2.4 Anionic Photoinitiators

Tertiary amine salts of ketocarboxylic acids[33] were used initially. Newer systems based on peptide chemistry have been described and used in microlithography.[34]

TABLE 5.5

Examples of Cationic Photoinitiators

5.2.5 Oxygen Inhibition of Cure

Oxygen decreases the efficiency of radical curing processes by quenching the triplet states of photoinitiators, scavenging radicals generated by the initiator system, and scavenging the growing macroradical.[35] In thin films, oxygen inhibition manifests itself by surface tack of cured films. To prevent this, the curing can be carried out under nitrogen blanket or by covering the film by a protective layer, such as a thin sheet of polyester. Another method is to incorporate a tertiary amine into the system. The latter method is preferred for UV curing[36]; however, it has a drawback, namely, an increased rate of photo-yellowing of cured films.[37]

5.2.6 Initiation of UV Hybrid Curing

Certain UV curing systems undergo polymerization by two different mechanisms, which is referred to as hybrid curing.[38] Photodecomposition of aryl-sulfonium salts generates both radicals and Brønsted acids. For example, the formed radicals can initiate the polymerization of acrylates, and the acid initiates the polymerization of vinyl ethers. In acrylate systems containing a cationic photoinitiator, the acrylate polymerization is greatly accelerated by the addition of vinyl ethers. On the other hand, the vinyl ether component is cured cationically. This produces interpenetrating polymer networks containing both the acrylate and vinyl ether component. Vinyl ethers as good hydrogen donors enhance photoinduced radical production from aryl sulfonium initiators, thus increasing the initiation rate of acrylate polymerization.[39]

In comparison with cationic curing systems, hybrid curing systems increase the rate of cure, produce cured films with improved solvent resistance, and offer a greater formulation latitude. When compared to free radical cure, better adhesion to critical substrates and lower oxygen sensitivity are observed in some cases.[40]

5.3 Kinetics of Photoinitiated Reactions

5.3.1 Kinetics of Free Radical Photopolymerization

Photoinitiated free radical polymerization proceeds via three main steps:

1. Initiation
2. Chain propagation
3. Termination

The initiation rate, v_i, depends on the radical yield per absorbed photon, Φ_i, and the number of photons absorbed per second in a unit volume, I_a. The latter quantity is a fraction of I_0, the number of photons per second per unit volume entering the process zone.

Initiation step:	AB*	$\rightarrow {}^\bullet R_1 + {}^\bullet R_2 \left({}^\bullet R_1, {}^\bullet R_2 - \text{free radicals} \right)$
	${}^\bullet R_1, {}^\bullet R_2 + M \rightarrow {}^\bullet P_1$	
Initiation rate:	$v_i = \Phi_i I_a$	

In the chain propagation, mainly the monomer is consumed and the propagation rate depends on the monomer concentration, [M], and the concentration of polymeric radicals, $[{}^{\bullet}P_n]$. The quantity k_p is the propagation rate constant.

Chain termination occurs by the combination or disproportionation of different polymer radicals. The termination rate, v_t, is proportional to the polymer radical concentration, $[{}^{\bullet}P_n]$, squared, with k_t being the termination rate constant. Other possible chain termination processes are chain transfer and reaction of polymer radicals with inhibitors and radical trapping (Andrzejewska, E., private communication).

Termination step: $^{\bullet}P_n + {}^{\bullet}P_m \rightarrow P_n - P_m$
                                                                or $P'_n - P_m$ (chain transfer)

Termination rate:

$$v_t = k_t \left[\cdot P_n \right]^2 \tag{5.3}$$

Since $v_i = v_t$, then

$$v_p = k_p / (k_t)^{\frac{1}{2}} [M] (\Phi_i I_a)^{\frac{1}{2}} \tag{5.4}$$

and thus

$$v_p \sim I_0 (1 - \exp(-2.303\,\varepsilon[PI]1))^{\frac{1}{2}} \tag{5.5}$$

The assumptions made to estimate the propagation rate, v_p, which is essentially the rate of the polymerization reaction, are:

1. The light used is monochromatic and is absorbed by the photoinitiator exclusively.
2. The absorption is small and homogeneous within the irradiated volume.
3. As the polymerization proceeds, a stationary radical concentration is obtained.
4. All polymer radicals exhibit the same reactivity toward propagation and termination.

5.3.2 Kinetics of Cationic Photopolymerization

Cationic polymerization is initiated either by strong Lewis acids, such as BF_3 or PF_5 or by Brønsted acids, such as H^+BF_4, H^+PF_6, or H^+SbF_6. Lewis acids are generated by UV irradiation of aryldiazonium salts, whereas diaryliodonium, triarylsulfonium, and triarylselenium salts produce upon UV irradiation strong Brønsted acids. The latter are preferred as initiating species in cationic polymerization.[36]

Reaction steps in a photoinduced cationic polymerization are as follows:

Initiation: $H^+X^- + M \rightarrow H-M^+ + X^-$

Propagation: $H-M^+ + nM \rightarrow H-(M)_nM^+$

Chain transfer: $H-(M)_nM^+ + ROH \rightarrow H-(M)_nM-OR + H^+$

Termination: $H-(M)_nM^+ + A^- \rightarrow H-(M)_nMA$

In the initiation step, the monomer, M, is initiated by intermediate protonation followed by the formation of a carbocation, $H-M^+$. Propagation can be terminated by anionic or nucleophilic species, A^-. If a hydroxy-functional compound (ROH) is present, chain transfer can occur via proton formation.

Photochemical decomposition of cationic photoinitiators can be sensitized by energy or electron transfer from the excited state of a sensitizer. Certain sensitizers, such as isopropylthioxantone, anthracene, or even certain dyes absorbing in the visible region, can be used to shift excitation to higher wavelengths. In such cases, visible light, for example, from tungsten lamps, can be used for photocuring.

There are several differences between the photoinitiated cationic polymerization and the photoinduced free radical polymerization:

1. The initiating species is a stable compound consumed only by anions and nucleophiles.
2. After the exposure to UV light, cationic polymerization continues for a long time.
3. Cationic systems are sensitive to humidity.
4. Cationic polymerization is not affected by oxygen the way the free radical polymerization is.
5. Cationic systems are generally slower than free radical curing systems.

The process is easily inhibited by trace amounts of basic materials, such as amines, urethanes, and certain fillers and pigments.[36]

Cationic curing gives a very low shrinkage and produces a more uniform and stable polymer. Coatings produced by cationic cure have excellent adhesion performance and chemical resistance, and are subject to thermal post-cure.[41]

A comprehensive review of the kinetics of photopolymerization of multifunctional monomers was published by Andrzcjcwska.[42]

5.4 Chemical Systems in UV Processing

The essential ingredients of any system processed by UV are:

- A reactive base oligomer, which imparts most of the properties to the final cured or cross-linked material; the molecular weight of oligomers used in UV processing is typically in the range from 400 to 7,000.
- Monofunctional monomer (or monomers) used to dilute the formulation to the suitable application viscosity.
- Multifunctional monomer(s) creating cross-links between segments of the oligomer and also acting as reactive diluent(s).[35]

There are many possibilities, but only a few systems are being used for practical UV curable formulations using free radical or cationic initiation. These will be discussed in this section.

5.4.1 Free Radical-Initiated Systems

5.4.1.1 Acrylate/Methacrylate Systems

The most widely used UV curable radical-initiated systems are based on acrylate unsaturation with the general formula $H_2C=CR-COOR'$ (if R=H, the monomer is an acrylate; if $R=CH_3$, it is a methacrylate). Methacrylates are less reactive than acrylates[43] but are less toxic and cause less skin irritation than acrylates.

The curing reaction of acrylates is typical of vinyl monomers. Therefore, the degree of double-bond conversion is the measure of the degree of cure.[43] The best results are obtained when using oligomers as binders and monomers as reactive thinners. Examples of difunctional and polyfunctional acrylates are in Table 5.6. A partial list of the most common acrylate oligomers is listed as follows.[13,44-46]

> *Epoxy acrylates:* The most widely used oligomers are aromatic and aliphatic epoxy acrylates. Epoxy acrylates are highly reactive and produce hard and chemically resistant films. The polymerization of monoacrylates produces linear polymers, whereas diacrylates produce branching and higher-functionality acrylates give rise to cross-linked structures. Cured materials are useful as coatings and adhesives on rigid substrates, such as metal cans and paneling, or as binders in composites. The epoxy component contributes to adhesion to nonporous substrates and enhances chemical resistance of the film.
>
> *Urethane acrylates:* Urethane acrylates are formed by the reaction of isocyanates with hydroxy-functional acrylate monomers. After UV

TABLE 5.6

Examples of Difunctional and Polyfunctional Acrylates Used in UV Curing

Monomer	Type of Main Chain	Functionality	Abbreviation	Molecular Weight	Viscosity, mPa.sec at 25°C
1,6-hexanediol diacrylate	Linear hydrocarbon	2	HDDA	226	5–10
Tripropyleneglycol diacrylate	Branched ether	2	TPGDA	300	10–20
Dipropyleneglycol diacrylate	Branched ether	2	DPGDA	242	8–12
Trimethylolpropane triacrylate	Branched hydrocarbon	3	TMPTA	296	50–150
Pentaerythritol triacrylate	Branched hydrocarbon	3	PETA	298	350–700
Ditrimethylolpropane tetraacrylate	Branched ether	4	DTMPTA	438	450–900
Dipentaerythritol pentaacrylate	Branched ether	5	DPEPA	525	14000

cure, they produce tough, flexible materials, which exhibit a good abrasion resistance.

Polyester acrylates: Acrylated polyesters are prepared by reacting the OH group of polyesters with acrylic acid or hydroxy acrylate with acid groups of the polyester structure. Polyester acrylates are often low-viscosity resins requiring little or no monomer.[47] They produce coatings and adhesives dominated by the polyester structure used in the oligomer. They are used for pressure-sensitive adhesives and also for strong rigid adhesives for metal-to-metal bonding. Amino-modified polyester acrylates show a high reactivity and low skin irritation.

Silicone acrylates: Acrylated organopolysiloxanes exhibit excellent release properties and are used as release coatings on papers and films. The silicone structure provides flexibility and resistance to heat, moisture, radiation degradation, and shear forces.[47]

Polyether acrylates: Polyether acrylates are produced by esterification of polyetherols with acrylic acid. They have a very low viscosity and do not require reactive thinners. Amino-modified polyether acrylates have a higher reactivity and low skin irritance, similar to polyester acrylates.

Solid urethane and polyester acrylates may be used as main components of radiation curable powders. Together with suitable unsaturated polyesters, powders are formed, which give low-film-flow temperatures and allow separating of film formation from curing. This technology has been used successfully in powder coating of wood and plastics.[48]

5.4.1.2 Styrene/Unsaturated Polyesters

Unsaturated polyesters have been among the earliest commercially available radiation curable systems. They are condensation products of organic diacids (or anhydrides) and glycols, most frequently using phthalic or maleic anhydride and 1,2 propylene glycol. In commercial systems, styrene is used as reactive diluent (at 20 to 40 wt. %) for unsaturated esters prepared from maleic anhydride and fumaric acid. The polymerization of styrene can be initiated by a wide range of thermal and Type I photoinitiators. Type II initiators cannot always be used because styrene quenches the triplet state of many carbonyl compounds.[49] The resulting product is styrene-polyester copolymer. The styrene/unsaturated polyester system is relatively slow but inexpensive and therefore has been used extensively for wood coatings, yet there is a tendency to replace them by acrylates.[50]

5.4.1.3 Vinyl Ether/Unsaturated Esters

When using a suitable Type I photoinitiator, for example, acylphosphine oxides and non-amine-containing acetophenones in this system, the polymerization process is very efficient and the resulting product is a 1:1 alternating copolymer.[51] This system is susceptible to oxygen inhibition but to a much lesser extent than acrylate polymerization.[52] An important advantage of this system is its low toxicity. Typical components include a variety of vinyl ethers and unsaturated esters, such as maleate, fumarate, citraconate, imides (maleimides), or N-vinylformamides.[53]

The system is also referred to as a maleate/vinyl ether (MA/VE) system because maleates are widely used in commercial products. The most widely and commercially used are systems consisting of oligomer components, which are not skin irritating or sensitizing, and exhibit a low vapor pressure. However, their cure speed is lower than that of acrylates. When radicals are generated by conventional photoinitiators, copolymerization occurs via an *electron donor-acceptor complex* (see Section 5.4.1.5). For practical applications, oligomeric MA/VA systems have been developed, consisting of maleate end caps and polyester, epoxy or urethane backbone section, which is diluted by means of a vinyl ether.[54]

5.4.1.4 Thiol-Ene Systems

This reaction is based on a stoichiometric reaction of multifunctional olefins (*enes*) with *thiols*. The addition reaction can be initiated thermally, photochemically, and by electron beam and radical or ionic mechanism.[55] Thiyl radicals can be generated by the reaction of an excited carbonyl compound (usually in its triplet state) with a thiol or via radicals, such as benzoyl radicals from a Type I photoinitiator, reacting with the thiol. The thiyl radicals add to olefins, and this is the basis of the polymerization process.[56] The addition of a dithiol to a diolefin yields linear polymer, higher-functionality thiols, and alkenes form cross-linked systems.

The most important attribute of the thiol-ene system is its insensitivity to oxygen; thus, it is not inhibited by it. Another attractive feature is its very high cure speed. The disadvantage of the thiol-ene system is an unpleasant odor of the volatiles emitted from some polythiol compounds.[50,56]

5.4.1.5 Donor-Acceptor Complexes

Another method to generate an initiating radical pair is by using donor-acceptor complexes (DACs). Unsaturated molecules containing an excess electron charge at the double bond—for example, vinyl ethers or styryloxy derivatives mixed with molecules that have electron-deficient unsaturation, such as maleic anhydride, maleates, or fumarates—form either *ground state* DAC (during mixing) or *excited state* DAC (after irradiation). Examples of such pairs are below[57]:

Methoxy styrene (donor)-maleic anhydride (acceptor)

Triethylene glycol divinyl ether (donor)-maleimide (acceptor)

1,4-Cyclohexane dimethanol divinyl ether (donor)-alkyl maleimide (acceptor)

The excited donor-acceptor complex undergoes hydrogen abstraction or electron transfer, thus generating the initiating radical pair for subsequent copolymerization. Some electron acceptors form strong chromophores for UV absorption, thus having the same function as a conventional photoinitiator, by being the source of free radicals. Such systems are potentially UV nonacrylate curing systems without photoinitiators.

Another development is free radical polymerization of neat acrylates without photoinitiator using 172 and 222 nm excimer lamps. Photons emitted from these radiation sources are absorbed by many acrylates generating free radicals directly.[58]

The development of self-curing resins, i.e., systems curing without photoinitiators or in some cases, with just small amounts of photoinitiators, has been reported.[59] Such resins are synthesized by Michael reaction of acrylic functional materials with Michael donor compounds such as acetoacetates. The resulting product has an increased molecular weight compared to the parent acrylate(s). This provides resins with reduced volatility and propensity for skin absorption. This new technology is versatile and flexible and opens a possibility of synthesis of a large number of different acrylate resins. The novel resins reportedly exhibit unique depth of cure capability. In the absence of a photoinitiator (PI), film of approximately 10 mils (0.25 mm) thick can be cured at a line speed of 100 fpm (30.5 m/min). When only 1% of PI is added, the thickness of film that can be cured increases to over 100 mils (2.5 mm).

5.4.2 Cationic Systems

As pointed out in Section 4.2.2, cationic polymerization processes are initiated by photoinitiators, which are essentially precursors generating Lewis and Brønsted acids. The mechanism of the process is ionic, and this chemistry does not function with the type of double bonds and unsaturation found in the monomers and oligomers reacting via free radical mechanism.

Cationic photoinitiation is based on the ring opening of the oxirane group. The photoinitiators of practical importance belong to three main classes of compounds: diazonium salts, onium (e.g., iodonium and sulfonium) salts, and organometallic complexes,[60] which upon irradiation by UV light decompose and yield an acid catalyst.

Aliphatic and cycloaliphatic epoxides (oxiranes) are by far the most frequently used substances in cationic process. However, cationic curing is not restricted to epoxies. Vinyl esters can also polymerize by a cationic mechanism and form an interpenetrating polymeric network when blended with epoxies. Multifunctional polyols are frequently used in cationic UV curing formulations as chain transfer agents, improving the cure speed, modifying the cross-link density, and affecting the flexibility of the coating.[61]

Epoxy-functional polydimethylsiloxane oligomers are another group that can be cured by UV radiation. Epoxysilicone block copolymers exhibit a good photoinitiator miscibility, high cure rate, and compatibility with epoxy and vinyl ether monomers. These block copolymers form flexible films with excellent release properties and are therefore used as release coatings.[54]

Cationically cured coatings exhibit a good adhesion to varied substrates, impact resistance, chemical resistance, and high hardness and abrasion resistance. The cure speed of cationic systems is generally lower than that of acrylates. Cationic cure is not affected by oxygen, but it is negatively affected by ambient humidity, amines, or other bases, and even by certain pigments. Heat increases the cure speed and often the ultimate degree of conversion. The acids generated by UV irradiation continue to be active even after the exposure stops. A pronounced dark postcure effect is observed and it may last several days.[54] An example of a general purpose cationic formulation is shown in Table 5.7.

TABLE 5.7

Example of a General-Purpose Cationic UV-Curable Formulation

Ingredient	Function	Weight %
Cycloaliphatic epoxide resin[a]	Base resin	70.4
ε-caprolactone triol	Flexibilizer and cross-linker	25.1
Arylsulfonium salts[b]	Photoinitiator	4.0
Silwet L-7604[c]	Silicone surfactant	0.5
Total		100.0

Source: Cyracure Cycloaliphatic Epoxies: Cationic UV Cure, Form 321-00013-0901 AMS, Dow Chemical Company, 2001.

[a] 3,4-epoxycyclohexylmethyl-3,4 epoxycyclohexane carboxylate.

[b] Mixed arylsulfonium hexafluorophosphate salts.

[c] OSi Specialities.

5.5 Photo-Cross-Linking of Polymers

While photopolymerization and photo-cross-linking of oligomers (or low-molecular-weight polymers) and monomers are well-established in industrial practice and widely used, a great deal of research has been conducted on the photo-cross-linking of high-molecular-weight polymers.[1] The energy sources for this process are high-intensity UV and visible light and lasers (1–10² eV), particularly in industrial applications.[62–64] The photoinitiators may be added to the polymer or grafted onto the polymer chains. Photocycloaddition involving poly(vinyl cinnamate), chalcone types of compounds, bismaleimides, antracene derivatives, cyclized polyisoprene,[65–67] and formation of cyclobutane ring by dimerization[68] and thiol-ene reactions based on polythiol and polyene (a polyunsaturated co-reactant) (see Section 5.4.1.4) using an aromatic ketone as photoinitiator[69] are examples of photo-cross-linking.

Because many polymers are to a certain degree photo-cross-linkable and photodegradable, the result of the overall reaction depends on the prevailing process, because cross-linking and degradation are competing processes.

The photo-cross-linkability of a polymer depends not only on its chemical structure but also on its molecular weight and the ordering of the polymer segments. Vinyl polymers such as PE, PP, polystyrene, polyacrylates, and PVC predominantly cross-link, whereas vinylidene polymers (polyisobutylene, poly-2-methylstyrene, polymethacrylates, and polyvinylidene chloride) tend to degrade. Likewise, polymers formed from diene monomers and linear condensation products, such as polyesters and polyamides, cross-link easily, whereas cellulose and cellulose derivatives degrade easily.[70]

There are several interesting applications reported in the literature, such as insulated wire and cable,[71,72] UV cross-linking of drawn fibers, and tapes from ultra-high-molecular-weight polyethylene.[73–75] Semi-interpenetrating networks (IPNs) from acrylates and polyurethanes are suitable as UV curable adhesives with high elasticity, good impact resistance, and excellent adhesion to a variety of substrates.[76,77]

The subject of photo-cross-linking is reviewed thoroughly by Rånby et al.[1]

References

1. Rånby, B., Qu, B. J., and Shi, W. F., in *Polymer Materials Encyclopedia* (Salamone, J. C., Ed.), CRC Press, Boca Raton, FL, p. 5155 (1996).
2. Calvert, J. G., and Pitts Jr., J. N., *Photochemistry*, John Wiley & Sons, New York (1966).
3. Birkus, J. B., *Photophysics of Aromatic Molecules*, John Wiley & Sons, New York (1970).

4. Birks, J. B., Ed., *Organic Molecular Photophysics*, Vols. 1 and 2, John Wiley & Sons, New York (1973, 1975).
5. Turro, N. J., *Modern Molecular Photochemistry*, Benjamin-Cummings, Menlo Park, CA (1978).
6. Klopfer, W., *Introduction to Polymer Spectroscopy*, Springer Verlag, Berlin, p. 27 (1984).
7. Holden, D. A., in *Encyclopedia of Polymer Science and Engineering*, Vol. 11 (Mark, H. F., and Kroschwitz, J. I., Eds.), John Wiley & Sons, New York, p. 128 (1988).
8. Jablonski, A., Z. *Physik*, 94, p. 38 (1935).
9. Wagner, P. J., and Hammond, G. S., *Adv. Photochem.*, 15, p. 21 (1968).
10. Guillet, J., *Polymer Photophysics and Photochemistry*, Cambridge University Press, Cambridge, UK (1985).
11. Davidson, R. S., *Exploring the Science, Technology and Applications of U.V. and E.B. Curing*, SITA Technology Ltd., London (1999).
12. Dietliker, K., in *Chemistry and Technology of UV and EB Formulations for Coatings, Inks and Paints*, Vol. 3 (Oldring, P. K. T., Ed.), SITA Technology Ltd., London (1991).
13. Fouassier, J. P., *Photoinitiation, Photopolymerization and Photocuring*, Hanser Publishers, Munich (1995).
14. Mehnert, R., Pincus, A., Janorsky, I., Stowe, R., and Berejka, A., *UV&EB Curing Technology & Equipment*, Vol. 1, John Wiley & Sons Ltd., Chichester/ SITA Technology Ltd., London, p. 15 (1998).
15. Pappas, S. P., in *Encyclopedia of Polymer Science and Engineering*, Vol. 11 (Mark, H. F., and Kroschwitz, J. I., Eds.), John Wiley & Sons, New York, p. 187 (1988).
16. Davidson, R. S., *Exploring the Science, Technology and Applications of U.V. and E.B. Curing*, SITA Technology Ltd., London, p. 67 (1999).
17. Timpe, H.-J., Miller, U., and Mockel, P., *Angew. Makromol. Chemie*, 189, p. 219 (1991).
18. Hageman, H. J., in *Photopolymerization and Photoimaging Science and Technology* (Allen, N. S., Ed.), Elsevier, Essex, UK (1989).
19. Davidson, R. S., in *Polymer Materials Encyclopedia* (Salamone, J. C., Ed.), CRC Press, Boca Raton, FL, p. 7330 (1996).
20. Herlihy, S. L., and Gatesby, G. C., in *Proceedings RadTech North America*, p. 156 (1994).
21. Gruber, G. W., U.S. Patents 4,017,296 (1977) and 4,024,296 (1977) to PPG Industries.
22. Ng, H. C., and Guillet, J. E., *Macromolecules*, 11, p. 937 (1978); Wistmontski-Kittel, T., and Kilp, T., *J. Polym. Sci. Polym. Chem. Ed.*, 21, p. 3209 (1983).
23. Gould, M. L., and Petry, V., "Photoinitiator and Cure Study," *PCT Paint & Coating Industry*, May 1, 2014.
24. Sitzmann, E. V., "Critical Photoinitiators for UV-LED Curing: Enabling 3D Printing, Inks and Coatings," *Proceedings, RadTech UV/EB West 2015*, Redondo Beach, CA, March 10–11, 2015.
25. Arcenaux, J. A., Wang, T., and Buomo, C., "UV LED Cure Applications," *Big Ideas Conference*, Redondo Beach, CA, March 10–11, 2019.
26. Mawby, T. R., "Pushing the Limits of LED Curing Forward to a Bright Future," *Proceedings, RadTech UV&EB Technology Expo and Conference, 2014*, Chicago, IL, May 12–14, 2014.
27. Greene, W. A., *Industrial Photoinitiators, Technical Guide*, Chapters 4 and 5, CRC Press, Boca Raton, FL (2010).

28. Wyrostek, M., and Salvi, M., "Photoinitiator Selection for LED-Cured Coatings," *UV+EB Technology*, 3, No. 2, p. 12 (2017).
29. Kiyoi, E., "The State of UV-LED Curing: Investigation of Chemistry and Applications," *UV+EB Technology*, RadTech ebook, RadTech International, North America, p. 60 (2016).
30. Plenderleight, R., "Photoinitiators for LED Curing: Development and Challenges for This Technology in the Future," *Proceedings, RadTech Europe Conference 2019*, Munich, Germany, October 15–16, 2009.
31. Pappas, S. P., in *Photopolymerization and Photoimaging Science and Technology* (Allen, N. S., Ed.), Elsevier, Essex, UK, Chapter 2 (1989).
32. Sahyun, M. R. V., DeVoe, R. J., and Olofson, P. M., in *Radiation Curing in Polymer Science and Technology*, Vol. II (Fouassier, J. P, and Rabek, J. F., Eds.), Elsevier, Essex, UK, Chapter 10 (1993).
33. Mayer, W., Rudolf, H., and DeCleur, E., *Angew. Makromol. Chem.*, 93, p. 83 (1981).
34. Cameron, J. F., and Frechet, J. M., *J. Organ. Chem.*, 55, p. 5919 (1990), and *Pure Appl. Chem.*, 61, p. 1239 (1991).
35. Davidson, R. S., in *Polymer Materials Encyclopedia* (Salamone, J. C., Ed.), CRC Press, Boca Raton, FL, p. 7334 (1996).
36. Hageman, H. J., in *Photopolymerization and Photoimaging Science and Technology* (Allen, N. S., Ed.), Elsevier, Essex, UK, Chapter 1 (1989).
37. Davidson, R. S., in *Polymer Materials Encyclopedia* (Salamone, J. C., Ed.), CRC Press, Boca Raton, FL, p. 7336 (1996).
38. Manus, P. J.-M., in *RadTech Europe Conference 1989*, Basel, Switzerland, p. 539 (1989).
39. Mehnert, R., Pincus, A., Janorsky, I., Stowe, R., and Berejka, A., *UV&EB Curing Technology & Equipment*, Vol. 1, John Wiley & Sons Ltd., Chichester/ SITA Technology Ltd., London, p. 21 (1998).
40. Crivello, J. V., *Adv. Polym. Sci.*, 62, p. 1 (1984).
41. Greene, W.A., *Industrial Photoinitiators, Technical Guide*, Chapter 7, CRC Press, Boca Raton, FL (2010).
42. Andrzejewska, E., *Progr. Polym. Sci.*, 26, p. 605 (2001).
43. Mehnert, R., Pincus, A., Janorsky, I., Stowe, R., and Berejka, A., *UV&EB Curing Technology & Equipment*, Vol. 1, John Wiley & Sons Ltd., Chichester/ SITA Technology Ltd., London, p. 2 (1998).
44. Elias, H. G., *An Introduction to Polymer Science*, VCH, Weinheim (1997).
45. Garratt, P. G., *Strahlenhartung*, Curt Vincentz Verlag, Hannover (1996).
46. Cray Valley Product Guide (Photocure), Cray Valley, Paris, www.crayvalley.com.
47. Mehnert, R., Pincus, A., Janorsky, I., Stowe, R., and Berejka, A., *UV&EB Curing Technology & Equipment*, Vol. 1, John Wiley & Sons Ltd., Chichester/ SITA Technology Ltd., London, p. 6 (1998).
48. Wittig, M., and Grohmann, T., in *Proceedings RadTech Europe '93*, p. 533 (1993).
49. Davidson, R. S., *Exploring the Science, Technology and Applications of U.V. and E.B. Curing*, SITA Technology Ltd., London, p. 42 (1999).
50. Mehnert, R., Pincus, A., Janorsky, I., Stowe, R., and Berejka, A., *UV&EB Curing Technology & Equipment*, Vol. 1, John Wiley & Sons Ltd., Chichester/ SITA Technology Ltd., London, p. 11 (1998).
51. Kohli, P., Scranton, A. B., and Blanchard, G. J., *Macromolecules*, 31, p. 568 (1998).
52. Hoyle, C., et al., ACS Symposium Series 673, p. 133 (1997).
53. Davidson, R. S., *Exploring the Science, Technology and Applications of U.V. and E.B. Curing*, SITA Technology Ltd., London, p. 46 (1999).

54. Mehnert, R., Pincus, A., Janorsky, I., Stowe, R., and Berejka, A., *UV&EB Curing Technology & Equipment*, Vol. 1, John Wiley & Sons Ltd., Chichester/ SITA Technology Ltd., London, p. 8 (1998).
55. Jacobine, A. F., in *Polymer Science and Technology*, Vol. III (Fouassier, J. P., and Rabek, J. F., Eds.), Elsevier Applied Science, London, p. 219 (1993).
56. Davidson, R. S., *Exploring the Science, Technology and Applications of U.V. and E.B. Curing*, SITA Technology Ltd., London, p. 43 (1999).
57. Mehnert, R., Pincus, A., Janorsky, I., Stowe, R., and Berejka, A., *UV&EB Curing Technology & Equipment*, Vol. 1, John Wiley & Sons Ltd., Chichester/ SITA Technology Ltd., London, p. 10 (1998).
58. Mehnert, R., Pincus, A., Janorsky, I., Stowe, R., and Berejka, A., *UV&EB Curing Technology & Equipment*, Vol. 1, John Wiley & Sons Ltd., Chichester/ SITA Technology Ltd., London, p. 19 (1998).
59. Sheridan, M., *Industrial Paint & Powder*, 78, No. 7, p. 12 (2002).
60. Fouassier, J.-P., *Photoinitiation, Photopolymerization and Photocuring*, Hanser Publishers, Munich, p. 102 (1995).
61. Mehnert, R., Pincus, A., Janorsky, I., Stowe, R., and Berejka, A., *UV&EB Curing Technology & Equipment*, Vol. 1, John Wiley & Sons Ltd., Chichester/ SITA Technology Ltd., London, p. 7 (1998).
62. Chen, Y. L., and Rånby, B., *J. Polym. Sci. Part A Polym. Chem.*, 27, p. 4051 (1989).
63. Chen, Y. L., and Rånby, B., *J. Polym. Sci. Part A Polym. Chem.*, 27, p. 4077 (1989).
64. Qu, B. J., Shi, W. F., and Rånby, B., *J. Photopolym. Sci. Technol.*, 2, p. 269 (1989).
65. Reiser A., *J. Chem. Phys.*, 77, p. 469 (1980).
66. Reiser A., *Photoreactive Polymers: The Science and Technology of Resists*, John Wiley & Sons, New York (1989).
67. Hasegawa, M., in *Comprehensive Polymer Science* (Allen, G., Ed.), Pergamon, Oxford, p. 1 (1989).
68. Fouassier, P. J., in *Radiation Curing in Polymer Science*, Vol. 1 (Fouassier, J. P., and Rabek, J. F., Eds.), Elsevier, London, p. 49 (1993).
69. Jacobine, A. F., in *Radiation Curing in Polymer Science*, Vol. 3 (Fouassier, J. P., and Rabek, J. F., Eds.), Elsevier, London, p. 219 (1993).
70. Rånby, B., Qu, B. J., and Shi, W. F., in *Polymer Materials Encyclopedia* (Salamone, J. C., Ed.), CRC Press, Boca Raton, FL, p. 5159 (1996).
71. Qu, B. J., et al., Chinese Patent 912 0815889 (1991).
72. Qu, B. J., et al., Chinese Patent Application (1991).
73. Chen, Y. L., and Rånby, B., *J. Polym. Sci. Part A Polym. Chem.*, 28, p. 1847 (1990).
74. Chen, Y. L., and Rånby, B., *Polym. Adv. Technol.*, 1, p. 103 (1990).
75. Zamotaev, P. V., and Chodak, I., *Angew. Makromol. Chem.*, 210, p. 119 (1993).
76. Moussa, K., and Decker, C., *J. Polym. Sci. Part A Polym. Chem.*, 31, p. 2633 (1993).
77. Moussa, K., and Decker, C., *J. Polym. Sci. Part A Polym. Chem.*, 31, p. 2197 (1993).

6

Electron Beam Processes

6.1 Introduction

As pointed out in Section 2.2, electrons, capable of exciting and ionizing molecules, have energies in the range of 5–10 eV. They are produced from fast electrons by energy degradation. As they penetrate solids and liquids, they generate ions, radicals, and excited molecules. The ionization results from inelastic collisions of fast electrons with the medium, and in the process the electrons lose energy. An empirical relationship expressing the energy of the electrons to the depth of electrons is due to Grun[1]:

$$R_G = 4.57E_0^{1.75} \qquad (6.1)$$

where R_G is the Grun range in pm and E_0 is the electron energy in keV.

This correlation is valid for a wide variety of materials, for polymers (e.g., polystyrene) and metals (e.g., aluminum). Thus, as the energy of the electrons increases, so does their penetration depth, and the amount of energy for high-energy electrons dissipated is small and constant over a large depth.[2] The stopping power of a material, which is the energy loss per unit path of an incident electron, depends on the density of the medium, and if it is a multicomponent system, it also depends on the relative concentrations of the individual components and on their molecular weights. This is important in cases where the organic medium contains pigments, which decelerates the incident electrons without yielding useful species.[3] However, the "decelerated" electrons react more readily with organic species.[4] The reactive species are dispersed randomly throughout the entire thickness of the material (see Chapter 2, Figure 2.5).

The energy deposited in the irradiated material causes a temperature rise (ΔT), which depends on absorbed dose and specific heat:

$$\Delta T = 0.239 \ D/c \qquad (6.2)$$

where D is the absorbed dose in kGy and c is the specific heat in cal/g °C of the irradiated material. Examples of temperature rise for different materials are shown in Table 6.1.

TABLE 6.1

Temperature Rise Values for Selected
Materials in Dependence on Absorbed
Dose

Material	ΔT °C/kGy
Polyethylene	0.43
Polypropylene	0.52
Polyvinyl chloride	0.75
Aluminum	1.11
Copper	2.63

Source: Drobny, J.G., Ionizing radiation and
         polymers, in *Principles, Technology
         and Applications*, Chapter 4, Elsevier,
         Oxford, UK, 2013.

   Radiation processing of monomers and polymers by electron beam, such
as polymerization and copolymerization of monomers, cross-linking, grafting, and degradation of polymers, is induced by these different chemically
reactive species.[5–7]

   Somewhat simplified, the process of interaction of high-energy electrons
with organic matter can be divided into three primary events[8]:

1. *Ionization:* In this event, the fast electron transfers its energy on
   the bonding electron in the absorbing material and the electron
   is knocked out. Ionization takes place only when the transferred
   energy during the interaction is higher than the bonding energy of
   the bonding electron.

$$AB \rightarrow AB^+ + e^-$$

   Practically, at the same time, the ionized molecule dissociates into a
   free radical and a radical ion:

$$AB^+ \rightarrow A\bullet + \bullet B^+$$

   The condition for further ionization is that the knocked-out electron
   still has enough energy to again ionize a molecule. If the molecules
   are hit by electrons that do not have sufficient energy for ionization,
   then excitation takes place.

2. *Excitation*: Excitation moves an electron from the ground state to the
   excited state.

$$AB \rightarrow AB^*$$

The excited molecule eventually dissociates into free radicals:

$$AB^* \rightarrow A\bullet + B\bullet$$

3. *Capture of electron:* This process is also ionization. Electrons with yet lower energy can be captured by molecules. The resulting ion can dissociate into a free radical and a radical ion:

$$AB + e^- \rightarrow AB^-$$
$$AB^- \rightarrow A\bullet + \bullet B^-$$

Besides these primary reactions, various secondary reactions arise in which ions or excited molecules take part.[9] The final result of these three events is that through the diverse primary and secondary fragmentations, radicals are formed that can initiate free radical processes, leading to polymerization, cross-linking, backbone or side-chain scissions, structural rearrangements, and other. The complete and rather eventful cascade of reactions triggered by the primary excitation of molecules may take up to several seconds. Under special conditions, such as when the polymer is in the glassy state or when the reactions occur in a crystalline matrix, transient species may survive for hours or even days. The energy deposited does not always cause change in the precise location where it is originally deposited, and it can migrate and affect considerably the product yield.[9] Migration of charges is another form of energy transfer in polymers, which can be electronic or ionic in nature. It can be either negative or positive and depends on the temperature.[10,11] Migration of radicals has been observed in irradiated polyethylene.[12]

Although radical cations are generated in some electron-irradiated monomers (e.g., vinyl ethers or epoxies), efficient cationic polymerization is not observed.[13] Under certain conditions (addition of iodonium, sulfonium, or sulfonium salts) cationic polymerization with the use of electron beam irradiation can be induced.[14] Several studies on radiation cross-linking of elastomers support the concept of ionic mechanism.[15–20]

In curing applications, electrons have to penetrate the reactive solids and liquids with typical masses per unit area of one to several hundred g/m^2 (1 g/m^2 = 1 μm at unit density). As shown in the depth-dose profiles in Figure 6.1,[9] electrons capable of producing any chemical change must have energies greater than 100 keV. Electrons with such energy can be generated by single-gap, low-energy electron accelerators. The *electron penetration range* is related to the path length and the electron travels during the energy degradation process. It can be estimated from the depth-dose distribution. Electron penetration range in g/m^2 as the function of electron energy is shown in Figure 6.2.

In industrial practice, materials processed by electron beam machinery such as coatings, rubber compounds, wire, and cable compounds often contain mineral fillers and other ingredients that increase the density of the material. Such change of density requires correction in order to assess the appropriate voltage for the given application.

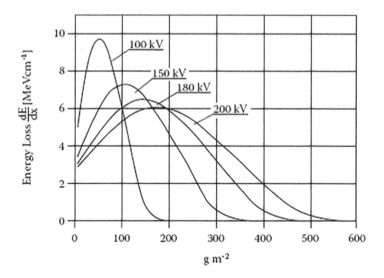

FIGURE 6.1
Tabata Ito Graph for electron penetration. (Modified from Tabata, T. and Ito, R., *Nucl. Sci. Eng.*, 53, 226, 1976.)

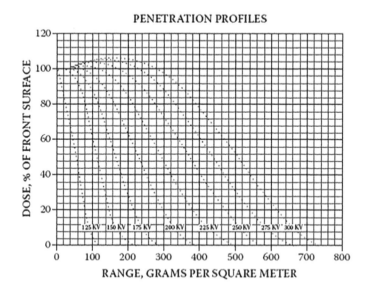

FIGURE 6.2
Electron penetration range in g/m^2 as a function of electron energy. (Courtesy Energy Sciences, Inc.)

The reactions initiated by the formation of radicals are the most important ones in the electron beam curing process. They can be inhibited by radical scavengers, such as oxygen, which, with its unpaired electrons, is an outstanding electron acceptor and consequently a radical inhibitor. It can readily inhibit polymerization reactions proceeding by the free radical mechanism typical mainly for acrylates and other vinyl monomers. This process is particularly pronounced on the surface of a coating that is being cured by electron beam. Therefore, in practical curing applications nitrogen blanketing is used to prevent direct contact of oxygen with the surface of the material being treated by the radiation.[21]

Although there are many changes in polymeric materials induced by electron beam irradiation, there will frequently be some that will control the net changes taking place. From the practical view, the two most important are *cross-linking* and *main chain scission* (i.e., polymer degradation), since they affect the physical properties of the product. Numerous studies attempting to find answers succeeded in explaining some correlations between the structure of a polymer and its response to ionizing irradiation, but there are still many unresolved issues. Several views have been advanced that explain some of the correlations. One study[22] suggests that when the monomeric unit of the polymer contains at least one α-hydrogen, cross-linking will take place—if not, the main chain will degrade. Another[23] proposed that vinyl polymers with two side chains attached to a single backbone carbon ($-CH_2-CR_1R_2-$) will degrade, and those with a single or no side chain ($-CH_2-CR_1H-$ or $-CH_2-CH_2-$) will cross-link. The correlation between the heat of polymerization and the tendency to degrade or cross-link was discussed by Wall.[24] Currently, the generally accepted mechanism involves the cleavage of a C–H bond on one polymer chain to form a hydrogen atom, followed by abstraction of a second hydrogen atom from a neighboring chain. The two hydrogen atoms form the H_2 molecule. Then the two adjacent polymeric radicals combine to form a cross-link leading to branched chains until, ultimately, a three-dimensional polymer network is formed when each polymer chain is linked to another chain. In contrast, scission is the opposite process of cross-linking in which the rupturing of a C–C bond occurs. Cross-linking increases the average molecular weight, whereas scission reduces it. If the energy of radiation is high, the chain breaks as a result of the cleavage of the C–C bond. In an oxygen atmosphere, however, the mechanistic way of scission proceeds in an indirect manner. The polymeric free radicals are generated by radiation, and the addition of oxygen to the polymeric radicals forms peroxidic species, which on decomposition form shorter chains.

It can be concluded that it is very difficult to predict the result from a polymer macrostructure, but it is relatively easy to measure the secondary species generated on irradiation by using known analytical techniques, such as measuring swelling, tensile tests, analysis using nuclear magnetic resonance (NMR), etc. The yield is then expressed by the *G value*, which represents the number of cross-links, scissions, double bonds, etc., produced for every 100 eV (1.6×10^{-17} J) dissipated in the material. For example, G (cross-links), abbreviated G(X), = 3.5 means that 3.5 cross-links are formed in the polymer per 100 eV under certain irradiation conditions. Similarly, the number of scissions formed is denoted by G(S). In order to determine the number of cross-links or G(X), the number of scissions or G(S), etc., it is necessary to know the dose or dose rate and the time of exposure for these irradiation conditions. From the product yields, it is possible to estimate what ratio of monomer units in a polymer is affected by irradiation.[25]

For cross-linked materials, changes in cross-link density are reflected by the extent to which the material is swelled while being immersed in a compatible solvent. If cross-linking predominates, the cross-link density increases and the extent of swelling decreases. If chain scission predominates, the opposite occurs. An increased soluble fraction then reflects scission, whereas reduced soluble fraction reflects cross-linking.

For an un-cross-linked material that undergoes predominantly cross-linking when exposed to ionizing radiation, solvent extraction experiments reveal that at certain absorbed dose (the critical dose), a percentage of the material is converted into insoluble gel. Beyond that point, the percentage of gel increases as a function of irradiation dose. In general, both the degrees of cross-linking and scissions can be determined from the soluble fraction by using the *Charlesby-Pinner equation*[26]:

$$s + s^{1/2} = p_0/q_0 + 10/q_0 D u_1 \qquad (6.3)$$

where s is the soluble fraction, p_0 is the density of scissions per unit dose, q_0 is the density of cross-links per unit dose, D is the absorbed dose in kGy, and u_1 is the number of average molecular weight. Then p_0/q_0 represents the ratio of chain scissions to cross-links.

The plot of $s + s^{1/2}$ vs. 1/D can be used to determine the p_0/q_0 ratio, and from that the values of G(S) and G(X) can be calculated.

Polymers can be classified into two groups according to their response to ionizing radiation. One group exhibits predominant cross-linking, the other predominant chain scission (see Table 6.2).[27]

Studies performed on elastomers have shown that the cross-linking efficiency is relatively low. G(X) values of 1 and approximately 3 were found

TABLE 6.2

Classification of Polymers According to Their Response to Ionizing Radiation

Polymers Predominantly Cross-Linking	Polymers Predominantly Degrading
Polyethylene	Poly(tetrafluoroethylene)
Polypropylene	Poly(α-methylstyrene)
Polystyrene	Poly(vinylidene chloride)
Poly(vinyl chloride)	Poly(vinyl fluoride)
Poly(vinyl alcohol)	Polychlorotrifluoroethylene
Poly(vinyl acetate)	Polytetrafluoroethylene
Poly(vinyl methyl ether)	Polyacrylonitrile
Polybutadiene	Polyvinylbutyral
Polychloroprene	Poly(methyl methacrylate)
Copolymer of styrene and acrylonitrile	Polymethacrylonitrile
Natural rubber	Polyoxymethylene
Chlorinated polyethylene	Poly(propylene sulfide)
Chlorosulfonated polyethylene	Poly(ethylene sulfide)
Polyamides	Cellulose
Polyesters	Polyalanine
Polyurethanes	Polylysine
Polysulfones	Polyisobutylene
Polyacrylates	DNA
Polyacrylamides	
Polydimethylsiloxane	
Polydimethylphenylsiloxane	
Phenol-formaldehyde resins	
Urea-formaldehyde resins	
Melamine-formaldehyde resins	

for natural rubber and polybutadiene, respectively, when irradiated at room temperature. Temperature was shown to have a positive effect in some cases.[28,29]

Experiments on cross-linking by irradiation have revealed that in many cases the cross-link densities equivalent to those obtained by conventional methods of cross-linking require relatively high radiation doses. Therefore, a great deal of experimental work has been done on the use of additives, which promote cross-linking by irradiation.[25] Such compounds are called radiation *cross-link promoters* or *prorads.*

6.2 Radiation Cross-Link Promoters

The cross-link promoters are prone to form more radicals faster than polymers alone, thus lowering radiation doses are required to achieve desired properties. There are essentially two types of promoters of radiation cross-linking or prorads.[30] One group, *indirect cross-link promoters,* which does not enter directly into the cross-linking reaction but merely enhances the formation of reactive species, such as free radicals, which then leads to secondary reactions to cross-linking. The other group, *direct cross-link promoters,* enters directly into the cross-linking reaction and becomes the actual connecting molecular links.

6.2.1 Indirect Cross-Link Promoters

6.2.1.1 Halogenated Compounds

This group of compounds has been studied extensively,[30–38] mainly in elastomers. The findings can be summarized as follows: For chlorinated aliphatic compounds, the sensitizing effect increases with decreasing number of carbon atoms in the molecule. The cross-link promoting effect increases on passing from iodo- to bromo- to chloro-substituted compound and with increasing degree of halogenation. Halogen-containing aromatic compounds also represent good cross-linking promoters.[39]

6.2.1.2 Nitrous Oxide

Enhanced radiation cross-linking in polyethylene, polypropylene, and polyisobutylene[40] and in copolymers of ethylene and propylene[41] was found when nitrous oxide was incorporated into the polymer matrix. Mechanisms of this process have been proposed by several workers.[39,42,43]

6.2.1.3 Sulfur Monochloride

Several investigators found a moderate[44] to substantial[45] increase in the rate of cross-linking of polyethylene and polypropylene. It was speculated that the differences between the results of the two studies might be due to the differences in the way the promoter was distributed in the test material.[46]

6.2.1.4 Bases

The promotion of cross-linking of polypropylene[47] and ethylene-propylene copolymers[44] has been confirmed. However, other bases, such as amines, were not found to be effective, and in some cases they acted as retarders of radiation cross-linking.[48]

6.2.2 Direct Cross-Link Promoters

6.2.2.1 Maleimides

Maleimides and dimaleimides are known to accelerate cross-linking of elastomers by organic peroxides at elevated temperatures,[49] but they also have been found to sensitize radiation-induced cross-linking of polymers.[50,51] Experiments involving radiation cross-linking of purified natural rubber and other elastomers have demonstrated that several maleimides, alkyl-, and aryldimaleimides significantly enhance the rate of cross-linking. Among the most effective were N-phenylmaleimide and m-phenylene dimaleimide, which in the amount of 5 wt. % increased the G(X) of purified natural rubber approximately by the factors of 23 and 15, respectively. The differences in the sensitizing action of the maleimides tested were at least attributed to their different solubilities in the rubber. For the m-phenylene dimaleimide, the rate of cross-linking was found to be directly proportional to its concentration up to 10 wt. %.[51]

The mechanism of the cross-link promotion of maleimides is assumed to be based on copolymerization of the polymer via its unsaturations with the maleimide molecules initiated by radicals, and in particular, by allylic radicals produced during the radiolysis of the polymer.[50]

When tested in other polymers, maleimides did not affect the rate of cross-linking in polydimethylsiloxane, polyisobutylene, and polyvinylchloride. In polyethylene, the addition of 5 wt. % of m-phenylene dimaleimide increased the G(X) from 1.8 to 7.2. In the polyvinylacetate, the effect was even more pronounced: the dose for gelation was reduced by about a factor of 50.[51] Contrary to the cross-link enhancing effect found for *m*-phenylene dimaleimide, cross-linking was retarded in polyvinyl acetate by the addition of monomaleimides. When analyzing the mechanism of the reaction, it was concluded that monomaleimides are not expected to affect cross-linking in saturated polymers.[52]

6.2.2.2 Thiols (Polymercaptans)

Studies of several investigators[53–58] have shown that polyfunctional thiol compounds are useful as cross-linking agents for the radiation curing of unsaturated elastomers and polymers used in graphic arts, electronics, and coating industries. Already small amounts of these compounds enhance the rate of cross-linking. They also promote cross-linking of polyisobutylene and its copolymers, which normally degrade on the exposure to radiation.[59]

The basis for cross-linking in the elastomers in these studies is the addition of thiols to olefinic double bonds. This reaction has been studied by a number of investigators.[55–59] It has been shown that the addition reaction proceeds by a radical mechanism. The rate of cross-linking of polybutadiene can be strongly promoted by relatively small amounts of polythiols.[60]

TABLE 6.3

Promotion of Cross-Linking by Polythiols

Elastomer	Prorad	Prorad Amount, wt%	G(X) Physical	Note
Polybutadiene	—	—	5	Irradiation in air (2.5 kGy/s)
	Dodecanethiol	1.0	4.5	Irradiation in air (2.5 kGy/s)
	Dimercaptodecane	1.0	29	Irradiation in air (2.5 kGy/s)
	Dimercaptodecane +2 wt.% o-dichlorobenzene	1.0	>49	
	Dipentene dimercaptan	1.0	18	
	α, α'-dimercapto-p-xylene	1.0	39	
	Trimethylolpropane trithioglycolate	1.0	>35	Irradiation in air (2.5 kGy/s)

For example, G(X) values increased from 5 (pure polybutadiene) to 29 when 1 wt.% of dimercaptodecane was added and to more than 49 on the further addition of o-dichlorobenzene (see Table 6.3).

6.2.2.3 Acrylic and Allylic Compounds

Polyacrylic and polyallylic compounds have been found to enhance the radiation cross-linking of polyvinylchloride.[61] They were found to also accelerate cross-linking of elastomers; however, their effect is rather small.[62] Radiation cross-linking of butadiene-styrene copolymers and natural rubber compounded with 50 phr (parts per hundred parts of rubber) of carbon black[63] in the presence of acrylate promoters exhibited greatly enhanced physical properties (modulus at 300% elongation and tensile strength). The best results were attained with tetramethyl diacrylate and ethylene dimethacrylate. Several polyfunctional allylic and acrylic compounds were evaluated in the radiation cross-linking of polyethylene, polypropylene, polyisobutylene, and ethylene-propylene copolymers. The rate of gelation of these polymers was increased considerably, but many of the radiation promoters homopolymerized on radiation exposure and, when compounded with elastomers, formed a rigid network structure within the more yielding elastomeric structure.[44,64,65] Examples of different polyfunctional monomers effective in a variety of polymers are given in Table 6.4.

TABLE 6.4

Examples of Effective Polyfunctional Prorads in Various Polymers

Polymers	Effective Polyfunctional Prorad(s)
Elastomers	
Styrene-butadiene copolymer (SBR)	Tetramethylol methane methacrylate
	Tetramethylene diacrylate
	Ethylene dimethacrylate
Chlorobutyl (CIIR)	Trimethylolpropane trimethacrylate
Polyisoprene rubber (IR)	Diethyleneglycol dimethacrylate
Polychloroprene (CR)	Polyethylene glycol dimethacrylate
Ethylene-propylene rubber (EPM)	Triethyleneglycol dimethacrylate
Ethylene-propylene-diene rubber (EPDM)	Ethylene glycol dimethacrylate
Butadiene-acrylonitrile rubber (NBR)	Diethyleneglycol dimethacrylate
Fluorocarbon elastomer (FKM)	Trimethylolpropane trimethacrylate
	Trimethylolpropane triacrylate
Natural rubber (NR)	Diethyleneglycol dimethacrylate
Conventional Plastics	
Polyethylene (PE)	Triallyl cyanurate
	Allyl methacrylate
Polypropylene (PP)	Trimethylolpropane trimethacrylate
Polyvinyl chloride (PVC)	Polyethylene glycol dimethacrylate
	Triallyl cyanurate
Polyvinylidene fluoride (PVDF)	Triallyl trially trimellitate
Ethylene-vinyl acetate copolymer (EVA)	Trimehylolpropane triacrylate
	Triallyl isocyanurate
Engineering Plastics	
Polyamide (PA), 6, 66,12	Triallyl isocyanurate
Polyamide (PA) 610	Triallyl isocyanurate
Polybutylene terephthalate (PBT)	Triallyl isocyanurate
Biodegradable Plastics	
Polycaprolactone (PCL)	Trimethallyl isocyanurate
Polybutylene succinate (PBS)	Trimethallyl isocyanurate

Source: Drobny, J.G., *Ionizing Radiation and Polymers-Principles, Technology, and Applications*, Chapter 5, Elsevier, Oxford, UK, 2013.

6.3 Retardants of Radiation Cross-Linking

Compared to the interest to promote the formation of cross-links, the protection of polymers against radiation damage has received much less attention. However, researchers discovered a large number of effective protective agents, often referred to as *antirads*, and also provided information on the yield of transient intermediates formed in the radiation process.[66] A considerable reduction of cross-linking was found for a number of aromatic amines, quinones, aromatic hydroxyl sulfur, and aromatic nitrogen compounds. The degree of protection offered by these compounds increases with their concentration but reaches a limiting value at a concentration of a few weight percent.[66]

The abovementioned radical acceptors are not the only effective protective agents against radiation. Studies performed with benzene and nitrobenzene in natural rubber[38] and styrene butadiene rubber (SBR) with N-phenyl-naphtylamine[67] demonstrated radiation protection of the respective compounds.

Many of the studied radical-accepting compounds were found to reduce the yield of scissions.[68,69] The antirads tested proved effective not only in the presence of air but also in its absence (see Table 6.5).

Different polymeric materials respond to irradiation by electron beam in different ways. A large number of them will be modified by the formation of a cross-linked network, by changing their surface properties or structure, and some will be degraded. Another field applicable to polymeric systems is polymerization and grafting. Electron beam can also be used for polymerization and cross-linking of oligomers and monomers, i.e., in conversion of liquids to solids.

TABLE 6.5

Effect of Antirads on the Yield of Cross-Links and Scissions in Natural Rubber

Natural Rubber Compound (50 phr EPC Carbon Black) + 5 phr Antirad	G(X)		G(S)[a]	
	Vacuum	Air	Nitrogen	Air
Control (1 phr phenyl-2-naphtylamine)	1.9	0.29	2.7	13
N-N'-dicyclohexyl-p-phenylenediamine	—	—	1.5	3
N-cyclohexyl-N'-phenyl-p-phenylene diamine	1.3	0.33	1.2	1.4
6-phenyl-2,2,4-trimethyl-1,2-dihydroquinoline	0.83	0.19	1.9	4.2
N-N'-dioctyl-p-phenylene diamine	0.87	0.12	1.5	5.0
2-naphtylamine	0.87	0.30	1.6	5.6
p-quinone	—	—	2.8	7.8
Phenylhydroquinone	1.1	0.46	2.2	5.4
1,4-naphtoquinone	1.1	0.48	2.0	5.6
2-naphtol	1.1	0.24	1.3	4.1
N-N'-diphenyl-p-phenylenediamine (35%) and phenyl-1-naphtylamine (65%)	0.97	0.27	1.4	3.7

[a] Scission yields are determined by stress relaxation measurements during irradiation of specimens. Permanent and temporary chain scissions are measured by this technique in contrast to gel measurements, which provide data only on permanent scissions.

6.4 Electron Beam Processing of Plastics

Plastics are by far the largest group of polymeric materials being processed by electron beam irradiation. Cross-linking of polyolefins, PVC, polyesters, polyurethanes, fluoropolymers, and fiber-reinforced composites is a common practice.

6.4.1 Polyolefins

Irradiation of polyolefins, particularly the family of polyethylenes, represents an important segment of the radiation processing. Polyolefins can be irradiated in many forms, such as pellets and powders, films, extruded and molded parts, or as wire and cable insulation.

Pellets and powders, as well as small parts, are handled simply by conveyor belts; cart systems or air transport tubes are the most common transportation systems for continuous electron beam irradiation.[70] Wire and cable as well as tubular products require special handling. Frequently, irradiation from one side is not sufficient because it results in nonuniform dose distribution. Therefore, the product has to be twisted so that its entire circumference is presented to the source of irradiation. Multiple passing is another way to alleviate this problem. Details are in Section 9.1 on wire and cable and Section 9.4 on heat-shrinkable products. Thick sheets and extruded profiles often have to be irradiated from both sides passing under the beam several times. Large and thick parts require spinning along some predetermined axis or angle to ensure uniform dose distribution.[70]

6.4.1.1 Polyethylene

The effects of ionizing radiation on polyethylene in all its forms can be summarized as follows:[71]

- The evolution of hydrogen
- The formation of carbon-carbon cross-links
- An increase in unsaturation to an equilibrium level
- A reduction in crystallinity
- The formation of color bodies in the resin
- Surface oxidation during irradiation in air

The formation of carbon-carbon cross-links is by far the most important effect and is the basis of the applications in the wire and cable industry and for heat-shrinkable products. The factors affecting the changes of polyethylene by irradiation are the molecular weight distribution, branching, degree of unsaturation, and morphology.[23]

Low-density polyethylene (LDPE), produced by a high-pressure polymerization, contains long branches attached to the main chain. Linear low-density polyethylene (LLDPE) has a rather regular succession of short-chain branches. Both types have a small amount of unsaturation, but high-density linear polyethylene (HDPE) contains one terminal vinyl group per molecule. At low radiation doses, this vinyl group has the effect of increasing the molecular weight of the HDPE by the chain formation of Y-links between the vinyl and the secondary alkyl radicals produced by the radiation.[72]

At ambient temperatures, polyethylene is always in the semicrystalline form. The amorphous region has a density of 800 kg/m³ (50 lb/ft³); the density of the crystalline region is 25% greater. The range of commercially available polyethylenes is from 920 to 960 kg/m³ (58 to 60 lb/ft³), but this relatively small difference in density corresponds to a considerably greater difference in the amorphous fraction: its weight fraction in LDPE is 40% and about 20% in HDPE.[73] The cross-linking takes place mainly in the amorphous region and the interface between the two phases. The cores of the crystalline regions sustain radiation-induced transvinylene formation in proportion to their weight fraction, but they are scarcely involved in gel formation.[73] Radiation cross-linking effects and the degree of cross-linking within crystalline regions are reported in the literature.[74–78]

Cross-linking of polyethylene by irradiation takes place normally at temperatures below 70°C (158°C), which is below its alpha transition temperature and much lower than the crystalline melting point. The irradiation does not have any significant effect on the sizes of amorphous and crystalline fractions as measured by the heats of fusion. However, when the cross-linked polymer is heated above the crystalline melting point, the crystallinity is significantly reduced because the cross-links interfere with the reformation of the supermolecular fraction during cooling from the melt. Subsequent cycles of melting and cooling do not produce additional changes in crystallinity.[73,78,79]

The core of the crystalline region of irradiated PE contains residual free radicals. These diffuse slowly to the interface with the amorphous region, where, in the presence of dissolved oxygen, whose equilibrium concentration is maintained by diffusion, they initiate auto-oxidative chain of degradation.[80] Postirradiation annealing in an inert atmosphere at a temperature above the alpha transition temperature (85°C or 185°F) leads to rapid mutual reactions of the free radicals and eliminates the problem.[81]

The higher crystalline fraction of the radiation cross-linked polyethylene even after a melt-freeze cycle has great technological merit for the heat-shrink packaging and electrical connector products.[82,83]

Compared with chemical cross-linking of PE, radiation curing produces a different product in many respects. The chemical cross-linking is done at temperatures near 125°C (257°F), where the polymer is in the molten state. Consequently, the cross-link density in the chemically cross-linked polyethylene is almost uniformly distributed, while there are relatively few cross-links in the crystalline fraction of the radiation cross-linked PE. The crystalline fraction of the radiation-processed polyethylene is greater than that in the chemically cured product.[73]

Radiation cross-linking of polyethylene requires considerably less overall energy and less space, and it is faster, more efficient, and environmentally more acceptable.[84] Chemically cross-linked PE contains chemicals, which are by-products of the curing system. These often have adverse effects on the dielectric properties and, in some cases, are simply not acceptable.[85]

The disadvantage of electron beam cross-linking is a more or less nonuniform dose distribution. This can happen particularly in thicker objects due to intrinsic dose-depth profiles of electron beams. Another problem can be a nonuniformity of rotation of cylindrical objects as they traverse a scanned electron beam. However, the mechanical properties often depend on the mean cross-link density.[86]

6.4.1.2 Polypropylene

Polypropylene (PP) is a stereospecific polymer prepared by polymerization using an organometallic catalyst system. Commercial polypropylenes have up to 95% isotactic content, which means that pendant methyl groups are almost all on the same side of the chain.

When polypropylene is exposed to ionizing radiation, free radicals are formed and these cause chemical changes. Because PP is highly crystalline, these radicals are relatively immobile and consequently may not be available for reaction for long periods of time.[87]

As with other polyolefins, upon irradiation the free radicals are formed along with evolution of hydrogen gas. If the radical is formed on the pendant methyl, the resulting reaction is cross-linking. However, if the radical is formed in the main chain, the chain end may react with hydrogen, thus causing an irreversible scission. Although the processes of chain scission and cross-linking occur simultaneously, and even though the net effect is cross-linking, the overall effect is the loss of mechanical strength.[88]

6.4.2 Polystyrene

Polystyrene is a clear, amorphous polymer with a high stiffness and good dielectric properties. It is easily cross-linked by ionizing radiation. Often, small amounts of divinyl benzene may enhance the degree of cross-linking.[89]

Moderate doses, between 1.0 and 20 Mrad (10–200 kGy) on partially or completely formed articles and polymer pellets, substantially reduce the content of residual monomer.[89] The mechanical properties of polystyrene are changed only at high radiation doses, which is characteristic of low cross-link yield and glassy morphology.[90]

6.4.3 Polyvinylchloride

Polyvinylchloride (PVC) is a slightly crystalline polymer because of its branched structure. A large amount of PVC is for insulations in the wire and cable industry. Radiation cross-linking of PVC requires the addition of a multifunctional prorad. In the absence of air, cross-linking predominates over chain scission and G(X) = 0.33. Addition of plasticizers or rising temperature above the glass transition increases cross-link yield.[91] Irradiated PVC materials usually show an increase in solvent, heat, and flow resistance.[92] Radiation curable PVC and its copolymers are suitable for grafting involving multifunctional monomers.[93]

The most effective prorads for PVC are acrylic and allylic esters, such as triallyl cyanurate (TAC), trimethylolpropane trimethacrylate (TMPTMA), and trimethylolpropane triacrylate. The triacrylate is more reactive than the trimethacrylate, but it is more toxic and, for that reason, is used only seldom. The amounts of these additives are 1% to 5% of the formulation weight.[94]

6.4.4 Polymethacrylates and Polyacrylates

Polymethyl methacrylate (PMMA) degrades under irradiation and becomes more soluble due to main chain scission.[95] The degradation can be greatly reduced by the addition of 10% of various additives, such as aniline, thiourea, or benzoquinone.[96] PMMA is an example of a nongelling polymer; it does not form a three-dimensional network structure under irradiation.[97]

6.4.5 Polyamides

In general, copolymers cross-link more readily than polyamide PA 66. Mechanical properties of polyamides are modified by irradiation, as seen by reduced tensile strength (50% loss when irradiated in air and 16% under vacuum). Aromatic polyamides retain strength better than aliphatic polyamides.[108]

6.4.6 Polyesters

Aliphatic polyesters tend to cross-link upon irradiation. Poly(ethylene terephtalate) cross-links with a low efficiency; however, it can sustain desirable physical properties up to 0.5 MGy.[108]

6.4.7 Fluoroplastics

6.4.7.1 Polytetrafluoroethylene

Under normal conditions, polytetrafluorothylene (PTFE) undergoes chain scission. In fact, this behavior is exploited commercially by converting PTFE scrap into a low-molecular-weight product that is then used in the form of very fine powders as additive to inks and lubricants. However, there is evidence that irradiation of PTFE at temperatures above its melting point (e.g., at 603–613 K, or 626–644°F) in vacuum results in a significant improvement in tensile strength and elongation at 473 K (392°F) and in the tensile modulus at room temperature.[98–100] These findings strongly contrast with the greatly reduced properties after irradiation at lower temperatures than that. This clearly indicates cross-linking in the molten state, similar to effects of irradiation on polyethylene. At temperatures above 623 K (662°F), thermal depolymerization is increasingly accelerated by irradiation and predominated over cross-linking at yet higher temperatures.[101]

When PTFE is degraded by ionizing radiation under vacuum, the amount of toxic compounds, such as HF and CO, decreases. The threshold value of absorbed dose for vacuum degradation of PTFE was found to be about 50–60 kGy.[102]

6.4.7.2 FEP

FEP, copolymer of tetrafluoroethylene (TFE) and hexafluoropropylene (HFP), has physical and chemical properties similar to those of PTFE, but it differs from it in that it can be processed by standard melt processing techniques.

When FEP is exposed to ionizing radiation at ambient temperature, it degrades like PTFE, with resulting reduction of physical properties. However, if the temperature of the polymer is raised before irradiation above its glass transition temperature, cross-linking predominates, and manifests itself by the increase in viscosity. With doses greater than 2.6 Mrad (26 kGy), ultimate elongation and resistance to deformation under load at elevated temperatures are improved, and the yield stress is increased. However, the improvements are offset by some loss in toughness.[103]

6.4.7.3 Other Fluoroplastics

Polychlorotrifluoroethylene (PCTFE), according to one source, degrades upon ionizing irradiation in a similar fashion as PTFE at ambient and elevated temperatures.[104] Unlike PTFE, when irradiated above its crystalline melting point, it still exhibits chain scission.[104] However, the resistance of PCTFE to ionizing radiation is reported to be better than that of other fluoropolymers.[104]

The copolymer of *ethylene and tetrafluoroethylene* (ETFE) can be cross-linked by irradiation.[105] Further improvement is achieved with the use of prorads, such as TAC or TAIC, in amounts up to 10%.

The copolymer of *ethylene and trichlorofluoroethylene* (ECTFE) behaves upon irradiation like ETFE, including improvement of cross-linking efficiency with prorads.

Polyvinylidene fluoride (PVDF) and *polyvinyl fluoride* (PVF) cross-link upon irradiation, particularly with the use of prorads, such as TAC, TIAC, diallyl itaconate, ethylene-bis-maleimide, and others.[104]

6.5 Electron Beam Processing of Elastomers

Elastomers are polymeric substances with rubber-like behavior at ambient temperatures. This means they are more or less elastic, extensible, and flexible. They can be extended by relatively small force and return to the original length (or near it) after the force is removed. Rubber-like behavior can be observed in plastics, but under different conditions, such as at elevated temperatures or in swollen state. These are not true elastomers, however.

The macromolecules of elastomers are very long and flexible, randomly agglomerated and entangled; the entangled macromolecular chains produce mechanical knots. Secondary forces acting between molecules are about a hundred times weaker than the primary bonds between the atoms connected within the molecule. The secondary forces are of physical nature, and their strength decreases rapidly with increasing distances between the molecules and with increasing temperature. The arrangement of the elastomeric molecules (flexibility and coiling of the chains, relatively weak intermolecular forces, chain entangling, and mechanical knots) gives elastomers the ability to undergo reversible deformations, but only under certain conditions: the deformation should not be too large and should last only a short time and within a narrow temperature interval. At higher deformations, particularly at elevated temperatures, the chains start to slide, the entanglements are reduced, and a permanent deformation takes place. The time dependence of these changes indicates a viscoelastic behavior of such polymeric material. The mechanical knots formed by the chain entanglements are not permanent, and the weak intermolecular forces cannot guarantee a sufficient shape stability of the material because they are affected by the changes of physical conditions (temperature, swelling) and they come into play only at low temperatures. Such material has only limited technical applications because it has poor mechanical properties and sensitivity to temperature changes. It is predominantly plastic and dissolves in some liquids to give colloidal solutions.

Only when chemical bonds between neighboring molecules are introduced is a raw elastomer converted into a rubber vulcanízate, which is essentially a three-dimensional network structure (see Figure 6.3). The process is referred to as vulcanization or curing, or more accurately, as cross-linking. A cross-linked elastomer, or rubber vulcanizate, is capable of large reversible deformations within a broad temperature range and does not dissolve but only swells in solvents and other liquids.

There are a large number of elastomeric materials used in commercial applications. They differ in the chemical nature of the monomer units, in their mutual arrangement and bonding, molecular weight, molecular weight distribution, branching, gel amount, etc. All these factors determine their chemical and physical properties, processing behavior, and solubility. Practically all commercial elastomers have a glass transition temperature below 0°C (32°F), and their molecular weights can range from 10,000 to 1,000,000.[106]

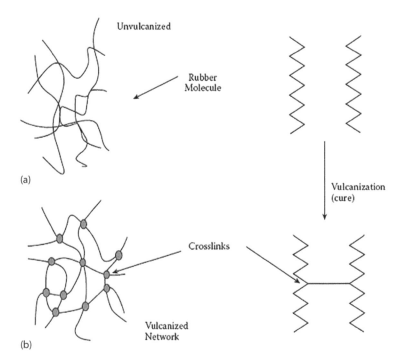

FIGURE 6.3
Elastomeric material. (a) uncross-linked, (b) cross-linked.

Vulcanization or cross-linking of elastomers is technically the most important process for conventional elastomers. During that process, strong chemical bonds are formed between molecules, restraining their mobility. As pointed out earlier, a three-dimensional network is formed. The cross-linking of elastomeric molecules is a random process; typically one cross-link is formed per 100–200 monomeric units.

The following changes occur as a result of cross-linking:

- The material changes from a plastic to elastic state.
- The product is much less sensitive to temperature changes.
- The material becomes stronger and often harder.
- The shape fixed by vulcanization (e.g., in a compression mold) cannot be changed unless it is subjected to mechanical work.

There are several methods of vulcanizing standard commercial elastomers. The classic method involves sulfur or sulfur-bearing compounds and peroxides, and these are typically used for hydrocarbon elastomers, such as natural rubber, styrene-butadiene rubber, polybutadiene rubber, EPM, and EPDM.

Other elastomers, such as polychloroprene, other chlorine-bearing elastomers, polyurethanes, and fluorinated rubber, use specific curing systems.
Details about curing methods, compounding, and processing are covered in
detail in several monographs.[107–110]

Regardless of the method of cross-linking, mechanical properties of a
cross-linked elastomer depend on cross-link density. Modulus and hardness increase monotonically with cross-link density, and at the same time,
the network becomes more elastic. Fracture properties, i.e., tensile and tear
strength, pass through a maximum as the cross-link density increases
(see Figure 6.4).

The cross-link density can be determined by equilibrium swelling or from
equilibrium stress-strain measurements at low strain rate, elevated temperature, and sometimes in the swollen state.[111]

As pointed out earlier, an elastomer cross-linked above its gel point will
not dissolve in a solvent, but will absorb it and swell. The swelling will continue until the forces of swelling balance the retroactive forces of the extended
chains of the network. The cross-link density can then be calculated from the
degree of swelling using the *Flory-Rehner equation*:

$$N = \frac{1}{2V_S} \frac{\ln\left(1-\phi+\phi+\chi\phi^2\right)}{\phi^{1/3}-\phi/2} \tag{6.4}$$

where N = moles of cross-links per unit volume (cross-link density),
V_s = molar volume of the solvent, ϕ = volume fraction of polymer in the
swollen gel, and χ = polymer-solvent interaction parameter.

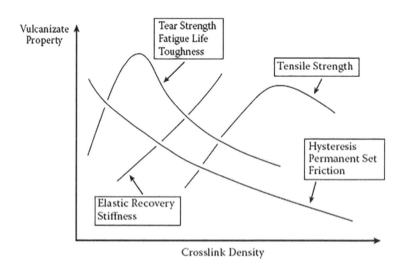

FIGURE 6.4
Effect of cross-link density on selected properties of an elastomeric material.

When using equilibrium stress-strain measurements, the cross-link density is determined from the *Mooney-Rivlin equation:*

$$\sigma/2(\lambda - \lambda - 2) = C_1 + C_2/\lambda \tag{6.5}$$

where σ = engineering stress (force per unit of original cross-sectional area), λ = extension ratio (measured length to original length), and C_1, C_2 = constants.

The value of C_1 is obtained from the plot of $\sigma/2(\lambda - \lambda^{-2})$ vs. $1/\lambda$ and extrapolating to $1/\lambda = 0$. By comparison with the theory of elasticity, it has been proposed that $C_1 = 1/2$ NRT, where N is the cross-link density, R is the gas constant, and T is the absolute temperature (of the measurement). To assure near-equilibrium response, stress-strain measurements are carried out at low strain rate, elevated temperature, and sometimes in the swollen state.[111]

6.5.1 Physical Properties of Radiation Cross-Linked Elastomers

Identical to chemically cross-linked (vulcanized) elastomers, the modulus of radiation-cured gum elastomers depends on the concentration of elastically effective network strands and temperature.[112]

Generally, values of tensile strength of radiation-cured elastomers are reported to be lower than those of sulfur-cured rubber.[113] It can be shown that main chain scissions, prevalent particularly under certain irradiation conditions, have a distinct effect on the strength of cured elastomers. The effect can be understood in terms of reduction in molecular weight and an associated increase in the chain-end contribution.[114] If a certain modulus is to be achieved in a vulcanizate, regardless of whether or not scission is occurring, the increase in chain ends (chains that have no load-bearing capacity) has to be compensated by an increase in the number of chemical cross-links. The associated decrease in the molecular weight of an elastically active strand (M_c) and its effect on failure properties such as energy to break were shown in Flory.[114] Since the maximum observable extension ratio $[(1/l_o)_{max}]$ is proportional to M_c, the increase in G(S)/G(X) encountered during radiation cure should decrease the $[(1/l_o)_{max}]$ and consequently cause lower failure properties, such as tensile strength, energy to break, etc.[115,116]

There are certain polymers, such as polyisobutylene, which by nature of their microstructure have a high yield of radiation-induced scissions, regardless of how they were irradiated. However, even polymers for which G(S) is negligible, when irradiated *in vacuo*, will undergo main chain scissions if the exposure is carried out at high temperature, in the presence of oxygen or certain additives. Degradation can also become significant with polymers having a low yield of cross-links G(X), since in the presence of a modest scission rate the ratio G(S)/G(X) can be of significant magnitude.[117] The compound ingredients, such as antioxidants, oils, etc., may retard the rate of cross-linking of an elastomeric compound. Thus, the presence of cure-retarding impurities and the exposure in an atmosphere containing oxygen may significantly contribute to the lower strength of radiation-cured elastomers. Another factor may be degradation of the polymer by ozone generated during irradiation.

Obviously, there are many subtle differences in the structure, morphology, or network topology between radiation-cured and sulfur-cured elastomers,[118] but their physical properties may be nearly equal, provided that precautions are taken to avoid the occurrence of chain scissions. A comparison of radiation cross-linked and sulfur-cured natural rubber (gum and carbon-black-reinforced compounds) is shown in Table 6.6.[119,120]

6.5.2 Effects of Radiation on Individual Elastomers

6.5.2.1 Natural Rubber (NR) and Synthetic Polyisoprene

Natural rubber and guttapercha consist essentially of polyisoprene in *cis*-1,4 and *frans*-1,4 isomers, respectively. Commercially produced synthetic polyisoprenes have more or less identical structures but reduced chain regularity, although some may contain certain proportions of 1,2 and 3,4 isomers. Microstructure differences not only cause the polymers to have different physical properties but also affect their response to radiation. The most apparent change in microstructure on irradiation is the decrease of unsaturation. It is further promoted by the addition of thiols and other compounds.[121] On the other hand, antioxidants and sulfur were found to reduce the rate of decay of unsaturation.[36] A significant loss in unsaturation was found particularly in polyisoprenes composed primarily of 1,2 and 3,4 isomers.[16,17]

A very important process occurring during irradiation is the formation of free radicals that leads to other changes, such as cross-linking, polymerization reactions, and grafting.

TABLE 6.6

Comparison of Stress-Strain Properties of Natural Rubber
Cross-Linked by Sulfur and by Radiation

Gum (Compound without fillers)	Sulfur-cured	Radiation-cured
Tensile strength, MPa (psi)	27.8 (4026)	18.6 (2700)
Elongation at break, %	700	760
300% modulus, MPa (psi)	2.3 (332)	1.7 (250)
Reinforced by 50 phr HAF Carbon Black	**Sulfur-cured**	**Radiation-cured**
Tensile strength, MPa (psi)	27.4 (3975)	22.8 (3300)
Elongation at break, %	470	350
300% modulus, MPa (psi)	14.8 (21.50)	17.9 (2600)

6.5.2.1.1 Gas Evolution

About 98% of gas formed during radiolysis of natural rubber and polyisoprene is hydrogen; the rest consists of methane and higher-molecular-weight
hydrocarbons. The yield of hydrogen is directly proportional to dose up to
200 Mrad (2,000 kGy). Moreover, it is independent of dose rate and the type
of radiation (gamma, electron beam).[38]

6.5.2.1.2 Cross-Linking

Cross-linking of natural rubber and synthetic polyisoprene has been studied
by several investigators using electron beam and gamma radiation.[122–124]
The general conclusion is that the yield of chemical cross-links, G(X), is
constant with dose and independent of dose rate and the type of radiation
used.[124] Effects of temperature are reported in Böhm and Tveekrem.[125]

The yield of cross-links in natural rubber and in high *cis*-polyisoprenes was
found to be about equal; however, polyisoprenes with high 1,2 and 3,4 configurations exhibited exceptionally high G(X). The presence of 1,4 units contributed
to the occurrence of chain scissions.[29,77] Oxygen has been found to increase
the rate of scission and to decrease the degree of cross-linking.[38,126] Another
interesting finding was that the rate of cross-linking is enhanced in crystalline
domains produced by strain orientation.[126] Maleimides and some halogenated
compounds enhance the cross-linking of natural rubber (see Section 6.2).

6.5.2.1.3 Physical Properties

The most commonly reported physical properties of radiation cross-linked
natural rubber and compounds made from it are modulus and tensile
strength, obtained from stress-strain measurements. Figure 6.5 illustrates

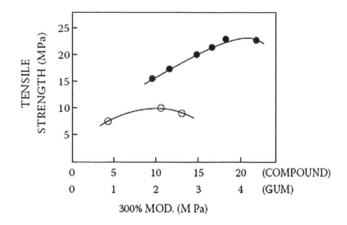

FIGURE 6.5
Tensile strength of radiation-cured purified natural rubber. (From Böhm, G.G.A. and Tveekrem,
J.O., *Rubber Chem. Technol.*, 55, 619, 1982.) Legend: ○ gum, ● compound (50 phr N330 carbon
black).

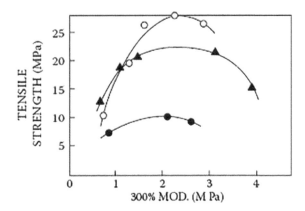

FIGURE 6.6
Tensile strength of radiation-cured purified natural rubber. (From Böhm, G.G.A., and Tveekrem, J. O., *Rubber Chem. Technol.*, 55, 620, 1982.) Legend: ○ sulfur, △ peroxide, ● EB irradiation in nitrogen at 2.5 kGy/s.

some of the results obtained from gum rubber and from a natural rubber compound reinforced by HAF carbon black.[126] In Figure 6.6, the tensile strength of radiation-cured gum is compared to that of vulcanizates cured by sulfur and by peroxide.[127]

Clearly, large doses are required for a full cure. The dose levels for attaining a maximum tensile strength are within 20–50 Mrad (200–500 kGy).

After exposure to such high levels of radiation, considerable modifications of natural rubber by isomerization and chain scission processes occur.[128] Because of that, the maximum strength values of gum rubber achieved by ionizing radiation are lower than those achieved by sulfur and peroxide curing systems. When prorads (cross-link promoters) such as dichlorobenzene, maleimides, or acrylates are used, the curing dose is reduced and higher tensile strength values are found. Tensile strength data from radiation-cured NR are shown in Table 6.7.

Somewhat greater tensile strength and elongation at break are obtained with high-energy electrons than with gamma radiation when exposure is carried out in air. However, when antioxidants are added to the compound, the differences diminish.[126,128]

The retention of the maximum tensile strength at elevated temperatures is greater for radiation-cured than for chemically cured natural rubber.[129–131] The physical properties after high-temperature aging are not improved, however. Lower flex life and higher abrasion resistance of radiation cross-linked NR were reported.[132] Other properties such as permanent set, hardness, and resilience were found to be nearly equal.

TABLE 6.7

Tensile Strength Data from Electron Beam-Cured Natural Rubber

Compound	Amount of Additive, phr	Dose, kGy	Tensile Strength, MPa (psi) Reported	Note
Gum rubber (SMR-5L) plus *o*-dichlorobenzene	3	140	9.0 (1305)	15 kGy/s
Gum rubber (Smoked sheets)	—	600	4.8 (696)	Irradiated in air
Smoked sheets and N330 Carbon black	50	500	19.6 (2842)	Irradiated in air
		140	13.3 (1928)	
Smoked sheets and N330 Carbon black plus octylacrylate	60	140	18.3 (2654)	Irradiated in air
	20			
Smoked sheets and N330 Carbon black plus tetramethylene diacrylate	55	140	18 (2291)	Irradiated in air
	10			
Smoked sheets and N330 Carbon black plus glyceryl triacrylate	50	130	16.6 (2407)	Irradiated in air
	5			
Pale crepe and N330 Carbon black	50	160	15.8 (2291)	Irradiated in air
Pale crepe and N330 Carbon black plus octylacrylate	50	160	16.6 (2407)	Irradiated in air
	20			

6.5.2.1.4 *Natural Latex*

The prevulcanization of natural rubber in latex form has also been a subject of much investigation.[133] The cross-linking mechanism is not yet fully understood, but the water apparently plays a major role in it. Irradiation results in the cross-linking of the rubber molecules and in coarsening of the latex particles. A process of cross-linking of natural rubber latex has been developed to the point that it can be used for an industrial-scale application.[133] The irradiation is performed in aqueous media by electron beam without a prorad (sensitizer) at a dose of 200 kGy (20 Mrad) or in the presence of *n*-butyl acrylate at considerably lower doses, typically 15 kGy. The cross-linked film exhibits physical properties comparable to those obtained from sulfur-cured (vulcanized) film. As an alternative, the addition of a variety of chloroalkanes makes it possible to achieve a maximum tensile strength with radiation doses of less than 5 Mrad (50 kGy).[133]

6.5.2.2 *Polybutadiene and Its Copolymers*

Homopolymers of polybutadiene can consist of three basic isomeric forms (*cis*-1,4, *trans*-1,4, and 1,2 vinyl), and these can be present in different sequential order. Copolymers may obtain a variety of comonomers, such as styrene, acrylonitrile, etc. Depending on their distribution in the chain, random copolymers or block copolymers of different types and perfection can be produced. There are many synthetic elastomers based on butadiene available commercially.

Upon irradiation, 1,4 polybutadienes and poly(butadiene-styrene) form free radicals relatively readily, and their concentration has been found to increase linearly proportional to dose up to approximately 100 Mrad (1,000 kGy).[134]

6.5.2.2.1 Gas Evolution

During radiolysis of polybutadiene and butadiene-styrene copolymers, hydrogen and methane evolve.[60,67,122,135–140] The incorporation of styrene as comonomer strongly reduces the total gas yield. Small amounts (typically 2 phr) of N-phenylβ-naphtylamine greatly reduce the gas yield[67] and at the same time reduce G(X) considerably.

6.5.2.2.2 Changes in Microstructure

Similar to NR, polybutadiene and its copolymers exhibit decay of unsaturation during radiolysis. Its extent depends greatly on microstructure.[138,139] Another effect observed is *cis-trans* isomerization.[140,141]

6.5.2.2.3 Cross-Linking

The yield of cross-links depends on the microstructure and purity of the polymer as well as whether it was irradiated in air or *in vacuo*.[131] The rate of degradation was found to be essentially zero when polybutadiene or poly(butadiene-styrene) was irradiated *in vacuo* but increased somewhat when irradiated in air.

The cross-linking of polybutadiene and its copolymers can be greatly enhanced by the addition of direct or indirect cross-link promoters. The largest G(X) values are obtained when using polymercaptans in combination with halogenated aromatics.[60,142–144] Carbon blacks and silica fillers were found to enhance cross-linking and, to some extent, become chemically linked to the polymer chain.[145,146] Some investigators suggest that electron beam irradiation of carbon-black-reinforced rubber compounds results in an increased cross-link density near the surface of the carbon black particle.[147] Aromatic oils enhance the occurrence of chain scissions and decrease the rate of cross-linking.[148] Common accelerators for rubber such as tetra-methyl-thiuram disulfide, sulfur, diphenylguanidine, and mercaptobenzo-thiazole inhibit radiation cross-linking, the degree of retardation being in the order of their listing.[138] As pointed out previously, several chemicals were reported to protect polybutadiene and its copolymers against radiation damage.

The reactions leading to the formation of cross-links between polymeric chains also can be employed to attach various compounds to the polymeric backbone. Using this technique, grafting of styrene,[149] mercaptans,[150] carbon tetrachloride,[150,151] and other compounds can be carried out.

6.5.2.2.4 *Physical Properties*

The information on physical properties of radiation cross-linking of polybutadiene rubber and butadiene copolymers was obtained in a fashion similar to that for NR, namely, by stress-strain measurements. From Table 6.8, it is evident that the dose required for a full cure of these elastomers is lower than that for natural rubber. The addition of prorads allows further reduction of the cure dose with the actual value depending on the microstructure and macrostructure of the polymer and also on the type and concentration of the compounding ingredients, such as oils, processing aids, and antioxidants in the compound. For example, solution-polymerized polybutadiene rubber usually requires lower doses than emulsion-polymerized rubber because it contains smaller amounts of impurities than the latter. Since the yield of scission G(S) is relatively small, particularly when oxygen is excluded, tensile strengths comparable to those attainable with standard sulfur-cured systems are obtained.[152]

Radiation-cured polybutadiene and its copolymers were reported to have superior tensile strength and ultimate elongation at high temperatures.[153] The abrasion resistance was found to be higher; most of the higher properties such as hardness, resilience, and permanent set were equal to those of chemically cured compounds of this category. Tensile strength values near equal to those obtained by chemically cured nitrite-butadiene rubber (NBR) were reported.[132,154,155] Relatively low radiation doses were required for a complete cure of these materials. Radiation-cured carbon-black-reinforced styrene-butadiene rubber (SBR) was found to have somewhat lower tensile strength and a lower maximum extensibility than sulfur-cured compounds, considerably longer fatigue flex life, and slower crack growth.[131]

6.5.2.3 *Polyisobutylene and Its Copolymers*

To this category belong homopolymers of polyisobutylene, copolymers of isobutylene and isoprene (butyl rubber), chlorobutyl, and bromobutyl. All these have been produced commercially for decades.

Upon irradiation, polyisobutylene and its copolymers tend to degrade. There are a large number of studies determining the nature of the process and mechanisms, and it was established that it involves formation and reaction of free radicals.[156] The radical concentration increases linearly with a dose up to 100 Mrad (1,000 kGy).[157]

6.5.2.3.1 *Gas Evolution*

Hydrogen and methane contribute to about 95% of the gas yield during radiolysis of polyisobutylene. The rest is made up of isobutylene and other mostly unidentified fragments.[158-162] It was found that the ratio of methane to hydrogen remains constant over a wide range of doses, whereas the ratio of isobutylene to hydrogen or methane rises rapidly with dose.[158,159] Moreover, it has been reported that the gas yields are largely independent of irradiation temperature.[163]

TABLE 6.8

Tensile Strength Data from Electron Beam Cross-Linked Polybutadiene and Its Copolymers

Polymer	Comonomer, mol%	Additive	Amount of Additive(s), phr	Dose, kGy	Tensile Strength, MPa (psi) Reported
Polybutadiene	0	N330 Carbon black and *p*-dichlorobenzene	50 2	80	15.2 (2204)
Poly(butadiene-styrene)	22.5	N330 Carbon black and *p*-dichlorobenzene	50 2	100	23.45 (3400)
	23.5	N330 Carbon black	50	280	20.4 (2958)
	25	N330 Carbon black	50	400	19.2 (2784)
		N330 Carbon black plus hexachloroethane	50 3	150	18.1 (2624)
Poly(butadiene-acrylonitrile)	32	Semi-reinforcing carbon black	50	100	22 (3190)

6.5.2.3.2 Changes in Microstructure

The main processes observed and reported are formation of vinylidene double bonds[158,159,164] and abstraction of methyl groups. The concentration of unsaturations located mainly at the end of a polymer chain produced during radiolysis was found to be increasing with dose.[163] Moreover, a linear relationship between unsaturations and scissions formed was noted over the temperature range of 83–363 K (–310°F–194°F). The ratio of double bonds to scissions was found to be independent of temperature and dose.[158,159,161]

6.5.2.3.3 Degradation and Cross-Linking

Degradation is the predominant process in irradiation of polyisobutylene. Generally, the scission yield increases with temperature in accord with the temperature dependence of the formation of unsaturation.[165] For poly(isobutylene-isoprene), it was found that the rate of degradation decreased with the increased content of isoprene in the copolymer, in accord with peroxide studies.[166] An extrapolation of this function to higher concentrations of unsaturation indicated that net cross-linking is expected to occur above approximately 5 mol% isoprene. A drastically different response to radiation was found for chlorinated poly(isobutylene-isoprene), poly(isobutylene-isoprene-divinylbenzene) terpolymers,[60] and dehydrohalogenated chlorobutyl rubber.[167] Rapid gelation occurred in those polymers already at low doses. The gradual depletion of these reactive sites by cross-linking then leads to prevailing degradation.[168] Certain additives can affect the response of these elastomers to irradiation. Polymercaptane compounds retard the net degradation of polyisobutylene and cause rapid gelation of poly(isobutylene-isoprene) and chlorobutyl rubber.[60,143,144] Allyl acrylate was found to cause gelation of polyisobutylene.[169]

6.5.2.3.4 Physical Properties

It is possible to cross-link polyisobutylene and poly(isobutylene-isoprene) and chlorobutyl rubber by means of certain additives; however, the physical properties of the vulcanizates prepared by conventional curing methods cannot be attained. The most promising results were obtained with chlorobutyl compounds containing thioether polythiols as prorads.[144]

6.5.2.4 Ethylene-Propylene Copolymers and Terpolymers

Commercial grades of ethylene-propylene copolymers (EPR) contain 60–75 mol% of ethylene to minimize crystallization. The addition of a third monomer such as 1,4-hexadiene, dicyclopentadiene, or 5-ethylidene-2-norbornene produces generally amorphous faster-curing elastomers. A large number of such terpolymers, referred to as EPDM, is available commercially. Their properties, performance, and response to radiation vary considerably depending on macrostructure, ethylene/propylene ratio, as well as on the type, amount, and distribution of the third monomer.

When EPDM is irradiated, the nature of the radical formed depends on the third monomer used[170,171] and, as in most cases, the radical concentration increases linearly with dose.[172]

6.5.2.4.1 Cross-Linking

The cross-linking rate of EPR by radiation comes close to that of polypropylene. EPDM terpolymers exhibit an enhanced cross-linking rate, and it increases with the diene content. However, not only the cross-linking rate but also a greater yield of scissions results from the addition of the third monomer.[44]

Cross-linking of EPR can be promoted by the addition of a variety of additives, particularly by those that were found effective in polypropylene. Tetravinyl silane, chlorobenzene, nitrous oxide, allyl acrylate, neopentyl chloride,[173,174] and N-phenyl maleimide[175] were reported to promote the process.

6.5.2.4.2 Physical Properties

Radiation-cured carbon-black-filled EPDM compounds exhibit tensile strength comparable to that of their analogues vulcanized by sulfur-accelerator systems.[176] Similar results were obtained from compounds containing up to 20 phr of acrylic additives, such as trimethylolpropane trimethacrylate, ethylene glycol dimethacrylate, triallyl phosphite, triallyl cyanurate, and others.[176] It should be noted, however, that some of the resulting compounds may be composed of two more or less independent, at least partially, separated networks. The noted lower compression set and reduced swelling in oils of the radiation-cured samples are most likely the result of this different morphology.[177] Among the variety of EPDM grades, those containing ethylidene norbornene exhibit the fastest rate of radiation cross-linking.[175,177,178]

Extender oils were found to cause a considerable increase in the dose required to attain the optimum cure. This can be explained by reaction of transient intermediates formed on the irradiated polymer chain with the oil and with the energy transfer, which is particularly effective when the oil contains aromatic groups. Thus, the ranking of oils as to their cure inhibition is aromatic > naphtenic > aliphatic.[177] This aspect is very important because many carbon-black-reinforced EPDM compounds frequently contain 100 phr or more oil.

6.5.2.5 Polychloroprene

Polychloroprene is the polymer of 2-chloro-1,3 butadiene. Emulsion polymerization produces an almost entirely *trans*-1,4 polymer, which is highly crystalline. Less crystalline polychloroprenes are produced by incorporating several wt.% of 2,3-dichloro-1,3 butadiene into the polymer to break up crystalline sequences.

Irradiation of carbon-black-reinforced polychloroprene compounds produced a maximum tensile strength of 20 MPa (2,900 psi) at a dose of 20 Mrad

(200 kGy), which is a value obtained typically from chemically cured compounds. The addition of 20 phr of *N,N'*-hexamethylene-bis methacrylamide as a prorad in the above compound produced a tensile strength of 18 MPa (2,610 psi) at a dose of 7 Mrad (70 kGy). Further addition of 6 phr of hexachloroethane caused the deterioration of the tensile strength by 50% at the 7 Mrad (70 kGy) dose.[179]

When irradiating a 1:1 blend of polychloroprene and poly(butadiene-acry-lonitrile) (NBR) reinforced by 50 phr furnace black and containing 5–15 phr of tetramethacrylate of bisglycerol phtalate, the product exhibited a tensile strength of 20 MPa (2,900 psi) at a dose of 15 Mrad (150 kGy) with values of elongation at break in the range of 420%–480%. These values are equal to or better than those obtained from similar compounds cured chemically.[180]

Irradiation of polychloroprene latexes of two different structures, one containing some sulfur and having a lower degree of branching and the other a highly branched polymer, made by mercaptane modification, showed a more rapid cross-linking of the branched polymer. The presence of the latex dispersion medium further enhanced the cross-linking process. The concentration of free radicals in the irradiated latexes was about 50% higher than that in the coagulated and subsequently dried rubber films.[181]

Polyorthoaminophenol (2 wt. %) and polyorthoaminophenol + phenyl-β-naphtylamine (0.7 wt. % each) acted as antirads, preventing any significant cross-linking on exposure to 24 krad (240 Gy).[182,183] The addition of a diphenylamine derivative gave reasonable protection to radiation up to 22 Mrad (220 kGy).[183]

6.5.2.6 Silicone Elastomers

There are a large variety of elastomers based on a chain –Si–O–Si– with different groups attached to the Si atom that affect the properties of the polymer. Technologically, the most widely used silicone elastomers are those with all methyls on the silicone atoms, i.e., polydimethylsiloxanes (PDMSs), or ones with less than 0.5 mol% of vinyl substitution for the methyls.[184]

Irradiation of PDMS produces hydrogen, methane, and ethane. The gas yield at room temperature correlates with the concentration of cross-links formed.[185] This can be expected since double bonds cannot be formed.

6.5.2.6.1 Cross-Linking

Cross-linking is a predominant process during irradiation of siloxane polymers. Chain scissions are negligible.[186–188] The cross-link density increases linearly with a dose up to 160 Mrad (1,600 kGy).[189] At 5.0 MGy (500 Mrad) the G(X) value is 0.5.[190] Free radical scavengers such as n-butyl and tert-dodecyl mercaptan and diethyl disulfide are the most effective antirads.[191,192] At a concentration of 10%, two-thirds of the cross-links were prevented from forming; however, the scission yield was also increased.

6.5.2.6.2 Physical Properties

The tensile strength of polydimethylsiloxane irradiated by a dose of 60 kGy (6 Mrad) is 15% lower than that of PDMS cross-linked by peroxide, but when 0.14 mol% of vinyl unsaturation is substituted for the methyl groups, the tensile strength is 30% higher.[190] Irradiation of PDMS containing 55 phr silica filler by a dose of 40 kGy exhibited a superior abrasion resistance when compared to peroxide vulcanizates even after prolonged thermo-oxidative aging at 523 K (482°F). The tensile properties of the materials prepared by both of these cross-linking methods were similar. An optimum tensile strength of 8 MPa (1,160 psi) was obtained on exposure to 4 Mrad (40 kGy).[190]

6.5.2.7 Fluorocarbon Elastomers

Fluorocarbon elastomers represent the largest group of fluoroelastomers. They have carbon-to-carbon linkages in the polymer backbone and a varied amount of fluorine in the molecule. In general, they may consist of several types of monomers: polyvinylidene fluoride (VDF), hexafluoropropylene (HFP), trifluorochloroethylene (CTFE), polytetrafluoroethylene (TFE), per-fluoromethylvinyl ether (PMVE), ethylene, or propylene.[193] Other types may contain other comonomers, for example, 1,2,3,3,3-pentafluoropropylene instead of HFP.[194] Fluorocarbon elastomers exhibit good chemical and thermal stability and good resistance to oxidation.

Since fluorocarbon elastomers being discussed here contain hydrogen in their molecules, they have the tendency to cross-link in addition to scission, common in fluoropolymers when exposed to radiation. The cross-linking predominates, but there is still a significant degree of chain scission.[195]

The use of cross-linking promoters (prorads)—such as diallyl maleate, triallyl cyanurate (TAC), triallyl isocyanurate (TAIC), trimethylolpropane methacrylate (TMPTM), and N, N'-(m-phenylene)-bismaleimide (MPBM) at amounts up to 10 wt.%—reduces the dose needed and the damage to the elastomeric chain by the radiation. It appears that individual fluorocarbon elastomers have the best cross-link yield with a specific prorad.[196–198] For example, copolymers of vinylidene fluoride (VDF) and hexafluoropropylene (CTFE) (e.g., Viton® A) and copolymers of vinylidene fluoride and chlorotrifluorethylene (Kel-F®) attain a high gel content and thermal stability up to 473 K (392°F) when cross-linked with a dose as low as 10–15 kGy without prorads, and only 2 kGy with 10 phr of multifunctional acrylates. Such thermal stability is unattainable with conventional chemical curing. Mechanical properties (tear resistance, tensile strength) are better than those from vulcanizates prepared by peroxide cross-linking.[199] The VDF-HFP copolymer is cross-linked by electron beam more efficiently than the copolymer of VDF and CTFE.[200]

Perfluoroelastomers (ASTM designation FFKM) are copolymers of two perfluorinated monomers, such as TFE and PMVE, with a cure site monomer (CSM), which is essential for cross-linking. Certain FFKMs can be cross-linked by ionizing radiation.[201] The advantage of radiation-cured FFKM is the absence of any additives, so that the product is very pure. The disadvantage is the relatively low upper-use temperature of the cured material, typically 150°C, which limits the material to special sealing applications only.[202]

6.5.2.8 Fluorosilicone Elastomers

Fluorosilicone elastomers generally respond to ionizing radiation in a fashion similar to that of silicone elastomers (polydimethylsiloxanes). One interesting application is a process of preparing blends of fluoroplastics, such as poly(vinylidene fluoride), with fluorosilicone elastomers to obtain materials having a unique combination of flexibility at low temperatures and high mechanical strength.[203]

Efficiencies of radiation-induced reactions of selected elastomers are shown in Table 6.9. Tensile strength data from another group of elastomers are shown in Table 6.10, and typical dose rates for cross-linking of selected elastomers are shown in Table 6.11.

TABLE 6.9

Efficiencies of Radiation-Induced Elastomer Reactions

Polymer	G(X)	G(H$_2$)[a]	G(S)/G(X)[a]
Polyisoprene (natural rubber)	0.9	0.43–0.67[b] 0.25–0.3[c]	0.16
Polybutadiene	3.6	0.23[d]	0.1–0.2
Styrene-bytadiene copolymer with 60 mol% styrene	0.6	0.11[d]	—
Styrene-butadiene copolymer with 23.4 mol% styrene	1.8–3.8	0.45 (87% H$_2$)	0.2–0.5
Ethylene-propylene rubber	0.26–0.5	3.3	0.36–0.54 G(S) = 0.3–0.46
EPDM (with dicyclopentadiene comonomer)	0.91	—	G(S) = 0.29
Polyisobutylene	0.05	1.3–1.6	G(S) = 1.5–5
Copolymer of vinylidene fluoride and hexafluoropropylene	1.7	0.27 G(HF) = 1.2	G(S) = 1.36
Copolymer of vinylidene fluoride and chlorotrifluoroethylene	1.03	—	G(S) = 1.56

Source: Drobny, J.G., *Ionizing Radiation and Polymers-Principles, Technology, and Applications*, Chapter 5, Elsevier, Oxford, UK, 2013.

[a] Unless otherwise stated.
[b] For doses up to 2 MGy (200 Mrad).
[c] Polyisoprenes with high 1,2 and 3,4 content.
[d] For a mixture of H$_2$ and CH$_4$.

TABLE 6.10

Tensile Strength Data from Selected EB-Irradiated Elastomers

Polymer	Additive	Amount of Additive, phr	Dose, kGy	Tensile Strength, MPa (psi)
Natural rubber	o-dichlorobenzene	3	140	9 (1305)
	Mono- and multi-functional acrylates	5–20	140	15–18 (2175–2610)
	Carbon black	50–60	1,000	1.6 (232)
	No additives	—	600	4.8 (696)
Polybutadiene	Carbon black and dichlorobenzene	50 2	8,000	15.2 (2204)
Copolymer of butadiene and styrene (23.5 mol % styrene)	Carbon black	50	28,000	20.4 (2958)
NBR (copolymer of butadiene and acrylonitrile) (32 mol% acrylonitrile)	Carbon black	50	10,000	22 (3190)
Polychloroprene	Carbon black	NA	20,000	20 (2900)
	Carbon black and diacrylamide	20 NA	700	18 (2610)
Blend of polychloroprene and NBR	Carbon black and multifunctional acrylate	50 5–15	15,000	20 (2900)

Source: Drobny, J.G., *Ionizing Radiation and Polymers-Principles, Technology, and Applications,* Chapter 5, Elsevier, Oxford, UK, 2013.

TABLE 6.11

Typical Dose Ranges for Cross-Linking of Selected Elastomers

Elastomer	Dose, kGy
EPM	50–150
EPDM	100–150
Chlorosulfonated PE ("Hypalon")	100–150
Vinyl-methyl silicone rubber	50–125

Source: Drobny, J.G., *Ionizing Radiation and Polymers*, Chapter 5, Elsevier, Oxford, UK, 2013.

6.5.2.9 *Thermoplastic Elastomers*

Thermoplastic elastomers (TPEs) are either block copolymers (SBS, SEBS, SEPS, TPU, COPA, and COPE) or blends, such as TPO (elastomer/hard thermoplastic, also referred to as thermoplastic olefin) and TPV (thermoplastic vulcanizate, blend of a vulcanized elastomer and a hard thermoplastic). These types represent the majority of the TPEs; other types are either specialty or small-volume materials.

Normally, TPEs are not cross-linked because the thermoplastic nature is their desired property in most cases. However, in some cases cross-linking is used to improve mechanical properties, influence flow[204] to reduce swelling in oils and solvents, eliminate dissolution of the polymer in oils and solvents, increase heat resistance, and influence other performance characteristics. Examples where cross-linking by ionizing radiation is necessary for the given processes are:

- Wire and cable insulation (to improve resistance to abrasion and cut-through).
- Thermoplastic foams (partial cross-linking of the material by EB increases the melt strength).
- Heat-shrinkable films, sheets, and tubing.

An example of cross-linking of copolyamides thermoplastic elastomer (COPA) by ionizing radiation to improve its heat resistance is shown in Table 6.12.

Thermoplastic elastomers based on polyolefins (TPO) are blends of PE or PP with EPDM elastomers wherein the elastomer is often cross-linked using thermochemical systems.[205] TPOs more suitable for medical products with no chemical residuals can be made using EB processing to cross-link the elastomer portion in such an elastomer-plastic blend. The thermoplastic governs the melt transition, and thus the extrusion properties of TPOs. The radiation response of these materials is also governed by the choice of the thermoplastic. An example of an EB-cured blend of EPDM and polyethylene used is for fluid transmission tubing and electrical insulation.[206]

A review article on electron beam processing of elastomers was written by Drobny.[207]

TABLE 6.12

Effect of Radiation Dose on Performance Characteristics of Copolyamide Formulation

Property	Dose, kGy				
	0	50	100	150	200
Tensile strength, MPa	59.5	65.7	51.4	42.9	Broke
Elongation at break, %	400	350	200	125	Broke
Hot creep elongation at 29 psi, 200°C, %	Melted	55	61	58	63

Source: Drobny, J.G., *Ionizing Radiation and Polymers-Principles, Technology, and Applications*, Chapter 5, Elsevier, Oxford, UK, 2013.

6.6 Electron Beam Processing of Liquid Systems

Electron beam processing of solvent-free liquid systems for coatings, inks, and paints involves essentially polymerization and cross-linking using electrons with energies between 120 and 300 keV. Initiation by electrons leads primarily to free radical reactions. Cationic polymerization is only found in rare cases.[208–212]

Typical liquid systems require doses between 10 and 50 kGy. They consist of binders (prepolymers) with acrylic ($H_2C=CH–CO–O–$) double bonds in the main chain (polymaleates and polyfumarates) and of monomers, usually acrylates used as reactive thinners.[212] Other ingredients added to the formulation may be pigments, dyes, fillers, flatting agents, and additives to improve film and surface properties and attain the required performance criteria.[213]

Reactive prepolymers used as binders are produced by acrylation of oligomers, such as epoxy resins, urethanes, polyesters, silicones, oligo-buta-diene, melamine derivatives, cellulose, and starches.[214–216] Prepolymers are the principal ingredients of coating formulations and it largely determines the basic properties of the coating. Examples of industrially important acrylated prepolymers are shown in Table 6.13.

Monomers, also called reactive thinners, are used to reduce the viscosity of the prepolymers but also have an effect on properties of the cured film. They form a high-molecular-weight network with the prepolymer after curing. In order to attain an adequate degree of cross-linking, principally bifunctional and polyfunctional acrylates are used. Monofunctional acrylates give a less reactive coating and are less desirable ingredients because of their volatility, odor, and skin-irritating effects. Currently, the following bifunctional and polyfunctional acrylates are used in industrial applications:

Tripropylene glycol diacrylate (TPGDA)

1,6-Hexanediol diacrylate (HDDA)

Dipropylene glycol diacrylate (DPGDA)

Trimethylolpropane triacrylate (TMPTA)

Trimethylolpropane ethoxytriacrylate (TMP(EO)TA)

Trimethylpropane propoxytriacrylate (TMP(PO)TA)

Pentaterythritol triacrylate (PETA)

Glyceryl propoxytriacrylate (GPTA)

Ethoxylated and propoxylated acrylates are highly reactive yet exhibit less skin irritation.[211]

TABLE 6.13

Examples of Acrylate Prepolymers Used for Radiation Technology

$$CH_2=CH-\overset{\overset{\displaystyle O}{\|}}{C}-O-(CH_2)_6-[O-(CH_2)_4-\overset{\overset{\displaystyle O}{\|}}{C}-O-(CH_2)_6-]-O-\overset{\overset{\displaystyle O}{\|}}{C}-CH=CH_2$$

Polyester acrylate

$$H_2C=HC-\overset{\overset{\displaystyle O}{\|}}{C}-O-CH_2\cdot\overset{\overset{\displaystyle OH}{|}}{CH}-R-\overset{\overset{\displaystyle OH}{|}}{CH}-CH_2\cdot O-CH=CH_2$$

R = Bisphenol A

Epoxy acrylate

$$CH_2=CH-\overset{\overset{\displaystyle O}{\|}}{C}-O-(CH_2)_6\cdot O-\overset{\overset{\displaystyle O}{\|}}{\underset{\underset{\displaystyle H}{|}}{C}}-N-R-N-\overset{\overset{\displaystyle O}{\|}}{\underset{\underset{\displaystyle H}{|}}{C}}-O-(CH_2)_6-O-\overset{\overset{\displaystyle O}{\|}}{C}-CH=CH_2$$

R = [structure with CH₃, NCO groups]   or   [structure with CH₃, NCO, NCO]   or   OCN−(CH₂)₆-NCO

Polyurethane acrylate

$$CH_2=CH-\overset{\overset{\displaystyle O}{\|}}{C}-O-R-\overset{\overset{\displaystyle CH_3}{|}}{\underset{\underset{\displaystyle CH_3}{|}}{Si}}-[O-\overset{\overset{\displaystyle CH_3}{|}}{\underset{\underset{\displaystyle CH_3}{|}}{Si}}-]_n O-\overset{\overset{\displaystyle CH_3}{|}}{\underset{\underset{\displaystyle CH_3}{|}}{Si}}-R'-O-\overset{\overset{\displaystyle O}{\|}}{C}-CH=CH_2$$

R, R' = $-CH_2-\overset{\overset{\displaystyle}{}}{\underset{\underset{\displaystyle}{|}}{C}}=CH_2$ or $-CH_2-CH=CH-$

Silicone acrylate

Electron beam curing of coatings requires an inert atmosphere to prevent oxygen inhibition. Inertization is done by nitrogen, obtained by vaporization of liquid nitrogen. Another advantage of using an inert atmosphere is to prevent formation of ozone, which is formed in the presence of air. Often, part of the gas is used for convective cooling of the accelerator beam window.

Many of the liquid systems curable by electron beam equipment are very similar to those cured by UV radiation. The EB process has the advantage that it offers the possibility of curing much thicker coats and pigmented formulations at much higher speeds than UV radiation. There is no need to use photoinitiator systems, which are often expensive and sometimes discolor the finished film. EB-cured coatings have frequently better adhesion to substrate because of the penetration of the electrons. These advantages are offset by the much higher capital cost of the EB equipment.[217] Another issue is to have enough production to sufficiently utilize the highly productive electron beam equipment. In some cases, to combine UV and EB processes is of technological or economical advantage.

6.7 Grafting and Other Polymer Modifications

6.7.1 Grafting

Radiation-induced grafting in its simplest form involves heterogeneous systems, with the substrate being film, fiber, or even powder and the monomer to be grafted onto the substrate is a neat liquid, vapor, or solution.[23,218–222] Currently, the three main radiation grafting techniques are [218,223]:

1. The pre-radiation procedure
2. The peroxidation procedure
3. Simultaneous method

In the *pre-radiation procedure*, the substrate is first irradiated, usually *in vacuo* or inert gas, to produce relatively stable free radicals, which are then reacted with a monomer, usually at elevated temperatures. A major advantage is minimization of homopolymer formation.[227]

The *peroxidation procedure*, which is the least often used of all the irradiation techniques, involves irradiation of the substrate in the presence of air or oxygen. This produces diperoxides and hydroperoxides on the surface of the substrate, which are stable, and the substrate can be stored until the combination with a monomer is possible. Monomer, with or without solvent, is then reacted with the activated peroxy trunk polymer in air or under vacuum at elevated temperatures to form the graft copolymer. The advantage of this method is the relatively long shelf life of the intermediate peroxy trunk polymers before the final grafting step.[224]

In the *simultaneous method*, which is the one most commonly used, the substrate is irradiated while in direct contact with the monomer. The monomer can be present as a vapor, liquid, or solution. This grafting process can occur via free radial or ionic mechanism.[225,226] With the simultaneous method, the formation of homopolymer is unavoidable, but there are several systems to minimize it. The advantage of this method is that both monomer and substrate are exposed to the radiation source and both form reactive sites. The other two techniques rely upon rupture of the bond to form reactive sites and therefore require higher radiation doses. Thus, the simultaneous method is more suitable for substrates sensitive to radiation. The simultaneous method can utilize UV radiation besides EB source. Logically, the UV irradiation requires a photoinitiator or sensitizer to achieve an acceptable level of grafting.

Radiation grafting can be performed with the monomers being neat or dissolved. In some cases, the use of solvents can produce graft copolymers with unique properties. Solvents, which wet and swell the backbone polymers, often assist grafting. Certain additives, including mineral acids and inorganic salts—such as lithium perchlorate, as well as monomers such as divinyl benzene (DVB) and trimethylolpropane triacrylate (TMPTA)—improve grafting yields.[223,227]

The dose rate affects both the yield and chain length of the grafted material. Air has a detrimental effect on grafting because it inhibits the reaction, which is consistent with other radiation-induced free radical reactions. Increasing the temperature of the grafting system increases the yield. This is very likely because raising the temperature increases the diffusion rate of the monomer into the substrate.[226]

Theoretically, radiation grafting is applicable to any organic backbone polymer. PVC grafting yields are among the highest under the radiation grafting conditions. Acid enhancement is observed for most backbone polymers. Solvent effects are also relevant to many backbone polymer systems (see Table 6.14). At low styrene solution concentrations (e.g., 30%) a peak in grafting is observed. Thus, the *Trommsdorff peak*,[228] is important because the chain length of graft is a maximum under these conditions and the graft yield reaches a maximum.

TABLE 6.14

Effect of Solvents on Radiation Grafting of Styrene to Polypropylene

	Grafting Yield, %							
	Styrene (wt.% in Solution)							
Solvent	20	30	40	50	60	70	80	100
Methanol	29	94	50	37	36	35	29	22
Ethanol	44	89	65	47	36	32	30	22
n-Butanol	123	74	34	40	33	29	28	22
n-Octanol	49	107	68	42	32	29	26	22
Dimethylformamide	24	40	43	44	40	39	33	22
Dimethylsulfoxide	11	29	66	61	56	42	24	21
Acetone	13	20	24	25	22	24	25	22
1,4-Dioxane	6	12	15	17	19	21	23	22

Source: Dworjanin, P. and Garnett, J.L., in *Radiation Processing of Polymers*, Singh, A. and Silverman, J., Eds., Carl Hanser Verlag, Munich, p. 98, 1992. With permission.

Note: Total dose 3 kGy, dose rate 0.4 kGy/h.

TABLE 6.15

Radiation Grafting of Miscellaneous Monomers[a]
onto Cellulose

Monomer	Grafting Yield, %
Styrene[b]	40
o-Methylstyrene[b]	110
p-Methylstyrene[b]	6
o-Chlorostyrene[b]	74
2-Vinyl pyridine[b]	3
Methyl methacrylate[c]	18
Vinylacetate[d]	11

Source: Dworjanin, P. and Garnett, J.L., in *Radiation Processing of Polymers*, Singh, A. and Silverman, J., Eds., Carl Hanser Verlag, Munich, p. 108, 1992. With permission.

[a] 40 vol.% in methanol.
[b] Total dose 5.4 kGy; dose rate 0.45 kGy/h.
[c] Total dose 10 kGy; dose rate 0.45 kGy/h.
[d] Total dose 10 kGy; dose rate 3.9 kGy/h.

Substituents have an effect as in conventional polymerization reactions. In radiation grafting, certain substituents activate monomers, and others deactivate them.[229,230] These effects can be seen in Table 6.15.

Multifunctional monomers such as acrylates (e.g., TMPTA) were found to have a dual function: to enhance the copolymerization and to cross-link the grafted trunk polymer chains. An addition of an acid along with a polyfunctional monomer has synergistic effects on grafting.[231]

Although many radiation-grafted materials have been discovered, only a limited number of them have been commercially utilized.[232,233] One of the first successful applications was the use of grafted films in battery separators.[234] Other possibilities are in ion exchange resins and membranes for separation processes.[235] The textile industry represents opportunities in improving flame retardation, permanent press, dyeing, and antistatic properties.[223,236] Other processes based on radiation grafting are useful in medicine (diagnostic and therapeutic) and industry (fermentation, bioseparation) and for producing catalyst supports.[237]

Radiation rapid curing (RRC) is another promising application of radiation grafting.[235] It is used for films containing oligomer-monomer mixtures, which cross-link within a fraction of a second when exposed to electrons from low-energy EB equipment.[238,239] In this process, multifunctional acrylates are used and have a dual function: they accelerate the rate of polymerization and also cross-link the film. The greatest potential is in packaging and in plastics industries for coating and ink applications.[237]

TABLE 6.16

Examples of EB-Induced Grafting for the Production of Separation Media

Polymer Substrate	Grafting Monomer/ Reaction	Application
Polyvinylidene fluoride film	Sodium styrene sulfonate	Fuel cell separation membrane
Polyvinylidene fluoride film	Acrylic acid and sodium styrene sulfonate	Improved hydrophilic membrane
Polyester/polyamide fabric	Acrylic acid	Recovery of copper and chromium from waste water
Polypropylene nonwoven fabric	Acrylonitrile	Uranium recovery from sea water

Source: Lapin, S.C., *UV+EB Technol.*, 1, 44, 2015.

The most recent review of grafting of polymers by EB irradiation or EB-induced graft copolymerization (EIGC) is by Lapin.[240] The application for resulting grafted copolymers discussed in this article include specialty fabrics, reinforcing fibers for fiber-reinforced composites (FRCs), plastics films with enhanced adhesion properties, and media used for separation and purification processes. Some known examples of the application of EIGC are shown in Table 6.16.

6.7.2 Other Polymer Modifications

Other polymer modifications involve surface or bulk modifications, and a majority of them are used in medical technology. Examples of these processes are:

- Modifications of surface of materials adapted for contact with human or animal tissues to impart biofunctional properties. These apply typically to ocular implants, surgical instruments, medical devices, or contact lenses.[241]
- Enhancing the wear resistance of polymers (e.g., ultra-high-molecular-weight polyethylene—UHMWPE—used for *in vivo* implants, such as artificial hip joints).[242]
- Electron beam cross-linking of hydrogels based on synthetic polyethylene oxide to produce materials for corneal prostheses.[243]
- Producing polyethylene microporous film with a porosity of 20%–80% for battery separators.[244]
- Three-dimensional processing of materials by electron beam to produce uniform isotropic irradiation of components, which are to be sterilized or bulk/surface modified.[245]

A comprehensive review of radiation techniques in the formulation of biomaterials was published by Kaetsu.[246] Substrate modification by electron beam was discussed by Wendrinski at the meeting of RadTech Europe 2001.[247]

References

1. Kase, K. R., and Nelson, W. R., *Concepts of Radiation Chemistry*, Pergamon Press, New York (1978).
2. Davidson, R. S., *Exploring the Science, Technology and Applications of U.V. and E.B. Curing*, SITA Technology, London, p. 123 (1999).
3. Lowe, C., in *Chemistry and Technology of U.V. and E.B. Formulations for Coatings, Inks and Paintings*, Vol. 4 (Oldring, P. K. T., Ed.), SITA Press, London, Chapter 1 (1991); Lowe, C., and Oldring, P. K. T., in *Chemistry and Technology of U.V. and E.B. Formulations for Coatings, Inks and Paintings*, Vol. 4 (Oldring, P. K. T., Ed.), SITA Press, London, Chapter 2 (1991).
4. Schafer, O., et al., *Photochem. Photobiol.* 50, p. 717 (1989).
5. Clegg, D.W., in *Irradiation Effects on Polymers* (Clegg, D. W., and Collier, A. A., Eds.), Chapter 1, Elsevier, London (1991).
6. Singh, A. and Silverman, J., in *Radiation Processing of Polymers* (Singh, A., and Silverman, J., Eds.), Chapter 1, Hanser Publishers, Munich (1992).
7. Mehnert, R., in *Ullmann's Encyclopedia of Industrial Chemistry*, Vol. A22 (Elvers, B., Hawkins, S., Russey, W., and Schulz, G., Eds.), VCH, Weinheim, p. 471 (1993).
8. Garratt, P., *Strahlenhartung*, Curt Vincentz Verlag, Hannover, p. 41 (1996). In German.
9. Tabata, T., and Ito, R., *Nucl. Sci. Eng.*, 53, p. 226 (1976).
10. Frankewich, E. L., *Usp. Khim.*, 35, p. 1161 (1966).
11. Hirsch, J., and Martin, E., *Solid State Commun.* 7, pp. 279 and 783 (1969).
12. Dole, M., Böhm, G. G. A., and Waterman, D. C., *Polym. J. Suppl.*, 5, p. 93 (1969).
13. Mehnert, R., Pincus, A., Janorsky, I., Stowe, R., and Berejka, A., *UV&EB Technology and Equipment*, Vol. 1, John Wiley & Sons, Chichster/SITA Technology Ltd., London, p. 24 (1998).
14. Buijsen, P., *Electron Beam Induced Cationic Polymerization with Onium Salts* (Thesis, Delft University of Technology), Delft University Press (1996).
15. von Raven, A., and Heusinger, H., *J. Polym. Sci.* 12, p. 2235 (1974).
16. Kaufmann, R., and Heusinger, H., *Makromol. Chem.* 177, p. 871 (1976).
17. Katzer, H., and Heusinger, H., *Makromol. Chem.* 163, p. 195 (1973).
18. Zott, H., and Heusinger, H., *Makromolekules*, 8, p. 182 (1975).
19. Kuzminskii, A. S., and Bolshakova, S. I., *Symposium Radiation Chemistry.* 3rd, Tihany, Hungary (1971).
20. Böhm, G. G. A., and Tveekrem, J. O., *Rubber Chem. Technol.* 55, p. 592 (1982).
21. Garratt, P., *Strahlenhartung*, Curt Vincentz Verlag, Hannover, p. 42 (1996). In German.
22. Miller, A. A., Lawton, E. J., and Balwit, J. S., *J. Polym. Sci.*, 14, p. 503 (1954).
23. Charlesby, A., *Atomic Radiation and Polymers*, Pergamon, Oxford (1960).
24. Wall, L. A., *J. Polym. Sci.*, 17, p. 141 (1955).

25. Böhm, G. G. A., and Tveekrem, J. O., *Rubber Chem. Technol.*, 55, p. 583 (1982).
26. Charlesby, A., and Pinner, S. H., *Proc. R. Soc. London Ser.*, A249, p. 367 (1959).
27. Clough, R., in *Encyclopedia of Polymer Science and Engineering*, Vol. 13 (Kroschwitz, J. I., Ed.), John Wiley & Sons, New York, p. 673 (1988).
28. Pearson, R. W., Bennett, J. V., and Mills, I. G., *Chem. Ind. (London)*, p. 1572 (1960).
29. Kozlov, V. T., Yevseyev, A. G., and Zubov, P. I., *Vysokomol. Soed.* A11, p. 2230 (1969).
30. Böhm, G. G. A., and Tveekrem, J. O., *Rubber Chem. Technol.*, 55, p. 593 (1982).
31. Chapiro, A. J., *Chim. Phys.*, 47, pp. 747 and 764 (1950).
32. Dogadkin, B. A., et al., *Vysokomol. Soedin.* 2, p. 259 (1960).
33. Jankowski, B., and Kroh, J., *J. Appl. Polym. Sci.*, 13, p. 1795 (1969).
34. Jankowski, B., and Kroh, J., *J. Appl. Polym. Sci.*, 9, p. 1363 (1965).
35. Kozlov, V. T., et al., *Vysokomol. Soedin.* A10, p. 987 (1968).
36. Kozlov, V. T., Klauzen, N. A., and Tarasova, Z. N., *Vysokomol. Soedin.* A10, p. 1949 (1968).
37. Nikolskii, V. G., Tochin, V. A., and Buben, N. Ya., Elementary Process of High Energy, *Izd. Nauka* (1965). In Russian.
38. Turner, D., *J. Polym. Sci.*, 27, p. 503 (1958).
39. Böhm, G. G. A., and Tveekrem, J. O., *Rubber Chem. Technol.*, 55, p. 596 (1982).
40. Okada, Y., *Adv. Chem. Ser.*, 66, p. 44 (1967).
41. Karpov, V. I., *Vysokomol. Soedin.*, 7, p. 1319 (1965).
42. Scholes, G., and Sinic, M., *Nature*, 202, p. 895 (1964).
43. Lyons, B. J., and Dole, M., *J. Phys. Chem.*, 68, p. 526 (1964).
44. Geymer, D. O., and Wagner, C. D., *Polym. Prep. Am. Chem. Soc. Div. Polym. Chem.*, 9, p. 235 (1968).
45. Kaurkova, G. K., Kachan, A. A., and Chervyntsova, L. L., *Vysokomol. Soedin.*, 7, p. 183 (1965); *J. Polym. Sci. Part C*, 16, p. 3041 (1967).
46. Böhm, G. G. A., and Tveekrem, J. O., *Rubber Chem. Technol.*, 55, p. 597 (1982).
47. Geymer, D., *Macromol. Chem.*, 100, p. 186 (1967).
48. Böhm, G. G. A., and Tveekrem, J. O., *Rubber Chem. Technol.*, 55, p. 600 (1982).
49. Miller, S. M., Roberts, R., and Vale, L. R., *J. Polym. Sci.*, 58, p. 737 (1962).
50. Miller, S. M., Spindler, M. W., and Vale, L. R., *Proc. IAEA Conf. Appl. Large Rad. Sources Ind.*, Salzburg, Vol. 1, p. 329 (1963).
51. Böhm, G. G. A., and Tveekrem, J. O., *Rubber Chem. Technol.*, 55, p. 601 (1982).
52. Böhm, G. G. A., and Tveekrem, J. O., *Rubber Chem. Technol.*, 55, p. 603 (1982).
53. Pearson, D. S., and Shurpik, A., U.S. Patent 3,848,502 to Firestone Tire and Rubber (1974).
54. Zapp, R. L., and Oswald, A. A., Paper 55, Meeting of the Rubber Division of the American Chemical Society, Cleveland (1975).
55. Griesbaum, K., *Angew. Chem. Ind. Ed. Engl.*, 9, p. 273 (1970).
56. Morgan, C. R., Magnotta, F., and Kelley, A. D., *J. Polym. Sci. Polym. Chem.*, 15, p. 627.
57. Morgan, C. R., and Kelley, A. D., *J. Polym. Sci. Polym. Lett.*, 16, p. 75 (1978).
58. Pierson, R. M., et al., *Rubber Plast. Age*, 38, pp. 592, 708, and 721 (1957).
59. Walling, C., and Helmreich, W., *J. Am. Chem. Soc.*, 81, p. 1144 (1978).
60. Böhm, G. G. A., *The Radiation Chemistry of Macromolecules*, Vol. II (Dole, M., Ed.), Academic Press, New York, Chapter 12 (1972).
61. Miller, A. A., *Ind. Eng. Chem.*, 51, p. 1271 (1959).
62. Böhm, G. G. A., and Tveekrem, J. O., *Rubber Chem. Technol.*, 55, p. 604 (1982).

63. Smith, W. V., and Simpson, V. G., U.S. Patent 3,084,115 to U.S. Rubber Co. (1963).
64. Lyons, B. J., *Nature*, 185, p. 604 (1960).
65. Odian, G., and Bernstein, B. S., *Nucleonics*, 21, p. 80 (1963).
66. Böhm, G. G. A., and Tveekrem, J. O., *Rubber Chem. Technol.*, 55, p. 606 (1982).
67. Blanchford, J., and Robertson, R. F., *J. Polym. Sci.*, 3, pp. 1289, 1303, 1313, and 1325 (1965).
68. Bauman, R. G., and Born, J. W., *J. Appl. Polym. Sci.*, 1, p. 351 (1959).
69. Bauman, R. G., *J. Appl. Polym. Sci.*, 2, p. 328 (1959).
70. Bradley, R., *Radiation Technology Handbook*, Marcel Dekker, New York, p. 100 (1984).
71. Bradley, R., *Radiation Technology Handbook*, Marcel Dekker, New York, p. 105 (1984).
72. Silverman, J., in *Radiation Processing of Polymers* (Singh, A., and Silverman, J., Eds.), Hanser, Munich, p. 16 (1992).
73. Silverman, J., in *Radiation Processing of Polymers* (Singh, A., and Silverman, J., Eds.), Hanser, Munich, p. 17 (1992).
74. Patel, G. N., and Keller, A., *J. Polym. Sci. Phys. Ed.*, 13, p. 305 (1975).
75. Chappas, W. J., and Silverman, J., *Radiat. Phys. Chem.*, 16, p. 437 (1980).
76. Liu, Z., Markovic, V., and Silverman, J., *Radiat. Phys. Chem.*, 25, p. 367 (1985).
77. Luo, Y., et al., *Radiat. Phys. Chem.*, 25, p. 359 (1985).
78. Lyons, B. J., *Radiat. Phys. Chem.*, 28, p. 149 (1986).
79. Zoepfl, F. J., Markovic, V., and Silverman, J., *J. Polym. Sci. Chem. Ed.*, 22, p. 2017 (1984).
80. Kashiwabara, H., and Seguchi, T., in *Radiation Processing of Polymers* (Singh, A., and Silverman, J., Eds.), Hanser, Munich, Chapter 11 (1992).
81. Silverman, J., in *Radiation Processing of Polymers* (Singh, A., and Silverman, J., Eds.), Hanser, Munich, p. 18 (1992).
82. Silverman, J., in *Radiation Processing of Polymers* (Singh, A., and Silverman, J., Eds.), Hanser, Munich, p. 19 (1992).
83. Silverman, J., in *Radiation Processing of Polymers* (Singh, A., and Silverman, J., Eds.), Hanser, Munich, p. 20 (1992).
84. Barlow, A., Biggs, J. W., and Meeks, L. A., *Radiat. Phys. Chem.*, 18, p. 267 (1981).
85. Fisher, P., *The Short-Time Electric Breakdown Behavior of Polyethylene*, Annual Report, National Academy of Science Publication, Washington, DC (1982).
86. Chappas, W. J., Mier, M. A., and Silverman, J., *Radiat. Phys. Chem.*, 20, p. 323 (1982).
87. Bradley, R., *Radiation Technology Handbook*, Marcel Dekker, New York, p. 114 (1984).
88. Bradley, R., *Radiation Technology Handbook*, Marcel Dekker, New York, p. 115 (1984).
89. Bradley, R., *Radiation Technology Handbook*, Marcel Dekker, New York, p. 132 (1984).
90. McGinniss, V. D., in *Encyclopedia of Polymer Science and Engineering*, Vol. 4 (Mark, H. F., and Kroschwitz, J. I., Eds.), John Wiley & Sons, New York, p. 418 (1986).
91. Wippler, C., *J. Polym. Sci.*, 24, p. 585 (1958).
92. Clough, R. L., and Gillen, K. T., *Radiat. Phys. Chem.*, 22, p. 527 (1983).
93. Posselt, K., *Kolloid. Z. Z. Polym.*, 223, p. 104 (1958).
94. Bradley, R., *Radiation Technology Handbook*, Marcel Dekker, New York, p. 134 (1984).

95. Charlesby, A., *Atomic Radiation and Polymers*, Pergamon Press, London, p. 336 (1962).
96. Alexander, P., Charlesby, A., and Ross, M., *Proc. Roy. Soc.*, 223, p. 392 (1954).
97. McGinniss, V. D., in *Encyclopedia of Polymer Science and Engineering*, Vol. 4 (Mark, H. F., and Kroschwitz, J. I., Eds.), John Wiley & Sons, New York, p. 442 (1986).
98. Tabata, Y., *Solid State Reactions in Radiation Chemistry, Taniguchi Conference*, Sapporo, Japan, p. 118 (1992).
99. Sun, J., Zhang, Y., and Zhong, X., *Polymer*, 35, p. 2881 (1994).
100. Oshima, A., et al., *Radiat. Phys. Chem.*, 45, p. 269 (1995).
101. Lyons, B. J., in *Modern Fluoropolymers* (Scheirs, J., Ed.), John Wiley & Sons Ltd., Chichester, p. 341 (1997).
102. Korenev, S., *Radiat. Phys. Chem.*, 71, p. 521 (2004).
103. Bowers, G. H., and Lovejoy, E. R., *I&EC Product Research and Development*, 1, June, p. 89 (1962).
104. Scheirs, J., in *Modern Fluoropolymers* (Scheirs, J., Ed.), John Wiley & Sons Ltd., Chichester, p. 61 (1997).
105. Carlson, D. P., and West, N. E., U.S. Patent 3,738,293, to E. I. du Pont de Nemours and Co. (1973).
106. Franta, I., in *Elastomers and Rubber Compounding Materials* (Franta, I., Ed.), SNTL Prague and Elsevier, Amsterdam, p. 25 (1989).
107. Quirk, R. P., and Morton, M., in *Science and Technology of Rubber*, 2nd ed. (Mark, J. E., Burak, E., and Eirich, F. R., Eds.), Academic Press, San Diego, Chapter 2 (1994).
108. Stephens, H.L., in *Handbook of Elastomers*, 2nd ed., Chapter 1 (Bhowmick, A., and Stephens, H. L., Eds.), Marcel Dekker, New York (2001).
109. Rothemeyer, F., and Sommer, F., *Kautschuktechnologie*, Carl Hanser Verlag, Munich (2001). In German.
110. White, J. L., *Rubber Processing*, Hanser Publishers, Munich (1995).
111. Hamed, G., in *Engineering with Rubber* (Gent, A. N., Ed.), Hanser Publishers, Munich, p. 20 (1992).
112. Böhm, G. G. A., and Tveekrem, J. O., *Rubber Chem. Technol.*, 55, p. 608 (1982).
113. Böhm, G. G. A., and Tveekrem, J. O., *Rubber Chem. Technol.*, 55, p. 610 (1982).
114. Flory, P. J., *Ind. Eng. Chem.*, 38, p. 417 (1946).
115. Böhm, G. G. A., et al., *J. Appl. Polym. Sci.*, 21, p. 3193 (1977).
116. Smith, T. L., in *Rheology, Theory and Application*, Vol. 5 (Eirich, F. R., Ed.), Academic Press, New York, Chapter 4 (1969).
117. Böhm, G. G. A., and Tveekrem, J. O., *Rubber Chem. Technol.*, 55, p. 611 (1982).
118. Böhm, G. G. A., and Tveekrem, J. O., *Rubber Chem. Technol.*, 55, p. 612 (1982).
119. Pearson, D. S., *Radiat. Phys. Chem.*, 18, p. 95 (1981).
120. Lal, J., *Rubber. Chem. Technol.*, 43, p. 664 (1970).
121. Kuzminskii, A., et al., in *Proceedings of the International Conference on Peaceful Uses of Atomic Energy*, Geneva, Vol. 29, p. 258 (1958).
122. Turner, D. T., *J. Polym. Sci.*, 35, p. 541 (1960); *Polymer (London)*, 1, p. 27 (1959).
123. Mullins, L., and Turner, D. T., *Nature*, 183, p. 1547 (1959).
124. Mullins, L., and Turner, D. T., *J. Polym. Sci.*, 43, p. 35 (1960).
125. Böhm, G. G. A., and Tveekrem, J. O., *Rubber Chem. Technol.*, 55, p. 617 (1982).
126. Roberts, D. S., and Mandelkern, L., *J. Am. Chem. Soc.*, 80, p. 1289 (1958).
127. Böhm, G. G. A., and Tveekrem, J. O., *Rubber Chem. Technol.*, 55, p. 619 (1982).
128. Harmon, D. J., *Rubber World*, 138, p. 585 (1958).

129. Harmon, D. J., *Rubber Age*, 86, p. 251 (1959).
130. Arnold, P. M., Kraus, G., and Anderson Jr., R. H., *Rubber Chem. Technol.*, 34, p. 263 (1961).
131. Dogadkin, B. A., Mladenov, I., and Tutorskii, I. A., *Vysokomol. Soedin.*, 2, p. 259 (1960).
132. Nablo, S. V., and Makuuchi, K., *Meeting of the Rubber Division of American Chemical Society* (October 11–14, Pittsburg, PA), Paper 83 (1994).
133. Chai, C. K., et al., *Sains Malaysiana*, 37, p. 79 (2008).
134. Böhm, G. G. A., and Tveekrem, J. O., *Rubber Chem. Technol.*, 55, p. 620 (1982).
135. Petrov, Ya., Karpov, V. L., and Vsesoyuz, T., *Soveshchenia po Radiatsionoi Khimii*, Academy of Science, USSR, p. 279 (1958).
136. von Raven, A., and Heusinger, H., *J. Polym. Sci.*, 12, p. 2255 (1974).
137. Kuzminskii, A., et al., in *Proceedings of the International Conference on Peaceful Uses of Atomic Energy*, Geneva, 29, p. 258 (1958).
138. Böhm, G. G. A., and Tveekrem, J. O., *Rubber Chem. Technol.*, 55, p. 622 (1982).
139. Parkinson, W. W., and Sears, W. C., *Adv. Chem. Ser.*, 66, p. 37 (1967).
140. Golub, M. A., *J. Am. Chem. Soc.*, 80, p. 1794 (1958).
141. Golub, M. A., *J. Am. Chem. Soc.*, 82, p. 5093 (1960).
142. Pearson, D. S., and Shurpik, A., U.S. Patent 3,843,502 to Firestone Tire & Rubber (1974).
143. Zapp, R. L., and Oswald, A. A., *Rubber Division Meeting 1975*, American Chemical Society, Cleveland, Paper 55 (1975).
144. Tarasova, Z. N., et al., *Kauch. Rezina*, 5, p. 14 (1958).
145. Harmon, D. J., *Rubber Age (London)*, 84, p. 469 (1958).
146. Böhm, G. G. A., et al., *J. Appl. Polym. Sci.*, 21, p. 3139 (1977).
147. Yoshida, K., et al., *J. Macromol. Sci.*, A14, p. 739 (1980).
148. Okamoto, H., Adachi, S., and Iwai, T., *J. Macromol. Sci.*, A11, p. 1949 (1977).
149. Okamoto, H., Adachi, S., and Iwai, T., *J. Polym. Sci.*, 17, pp. 1267 and 1279 (1979).
150. Böhm, G. G. A., and Tveekrem, J. O., *Rubber Chem. Technol.*, 55, p. 629 (1982).
151. Dogadkin, B. A., et al., in *Proceedings of the All-Union Conference*, Riga (1960).
152. Jackson, W. W., and Hale, D., *Rubber Age*, 77, p. 865 (1955).
153. Pearson, D. S., and Böhm, G. G. A., *Rubber Chem. Technol.*, 45, p. 193 (1972).
154. Böhm, G. G. A., et al., *J. Appl. Polym. Sci.*, 21, p. 3193 (1977).
155. Böhm, G. G. A., and Tveekrem, J. O., *Rubber Chem. Technol.*, 55, p. 634 (1982).
156. Palalskii, B. K., et al., *Vysokomol. Soedin.*, A16, p. 2762 (1974).
157. Alexander, P., and Charlesby, A., *Proc. R. Soc. London*, A230, p. 136 (1955).
158. Alexander, P., Black, R., and Charlesby, A., *Proc. R. Soc. London*, A232, p. 31 (1955).
159. Berzkin, B. G., et al., *Vysokomol. Soedin.*, 9, p. 2566 (1967).
160. Turner, D. T., in *The Chemistry and Physics of Rubber Like Substances* (Bateman, L., Ed.), John Wiley & Sons, New York, p. 563 (1963).
161. Karpov, V. L., *Sessiya Akad. Nauk SSSR po Mirnomu Ispol. Atom. Energii*, Akad. of Sci., Moscow (1955).
162. Böhm, G. G. A., and Tveekrem, J. O., *Rubber Chem. Technol.*, 55, p. 635 (1982).
163. Slovokhotova, N. H., and Vsesoyuz, I. I., *Seveshchania po Radiatsionoi Khimii*, Academy of Sciences, Moscow, p. 263 (1958).
164. Böhm, G. G. A., and Tveekrem, J. O., *Rubber Chem. Technol.*, 55, p. 637 (1982).
165. Loan, L. D., *J. Polym. Sci.*, A2, p. 2127 (1964).
166. EXXON chemical brochure on conjugated diene butyl elastomer.

167. Böhm, G. G. A., and Tveekrem, J. O., *Rubber Chem. Technol.*, 55, p. 638 (1982).
168. Odian, G., and Bernstein, B. S., *J. Polym. Sci. Lett.*, 2, p. 819 (1964).
169. Geissler, W., Zott, M., and Heusinger, H., *Makromol. Chem.*, 179, p. 697 (1978).
170. Faucitano, A., Martinotti, F., and Buttafava, A., *Eur. Polym. J.*, 12, p. 467 (1969).
171. Böhm, G. G. A., and Tveekrem, J. O., *Rubber Chem. Technol.*, 55, p. 640 (1982).
172. Odian, G., Lamparella, D., and Canamare, J., *J. Polym. Sci.*, C16, p. 3619 (1967).
173. Smetanina, L. B., et al., *Vysokomol. Soedin.*, A12, p. 2401 (1970).
174. Pearson, D. S., and Böhm, G. G. A., *Rubber Chem. Technol.*, 45, p. 193 (1968).
175. Eldred, R. J., *Rubber Chem. Technol.*, 47, p. 924 (1974).
176. Böhm, G. G. A., and Tveekrem, J. O., *Rubber Chem. Technol.*, 55, p. 642 (1982).
177. Kammel, G., and Wiedenmann, R., *Siemens Forsch. Entwicklungsber.*, 5, p. 157 (1976).
178. El. Miligy, A. A., et al., *Elastomerics*, 111, p. 28 (1979).
179. Krishtal', I. V., et al., *Kauch. Rezina*, 4, p. 11 (1976).
180. Tarasova, N. N., et al., *Vysokomol. Soedin.*, A14, p. 1782 (1972).
181. Oganesyan, R. A., *Kauch. Rezina*, 1, p. 34 (1976).
182. Ito, M., Okada, S., and Kuriyama, I., *J. Mater. Sci.*, 16, p. 10 (1981).
183. Polmanteer, K. E., in *Handbook of Elastomers* (Bhowmick, A. K., and Stephens, H. L., Eds.), Marcel Dekker, New York, p. 606 (2001).
184. Ormerod, M. G., and Charlesby, A., *Polymer*, 4, p. 459 (1963).
185. Charlesby, A., *Proc. R. Soc. London Ser. A*, 230, p. 120 (1955).
186. Miller, A. A., *J. Am. Chem. Soc.*, 82, p. 3319 (1960).
187. Kilb, R. W., *J. Phys. Chem.*, 63, p. 1838 (1959).
188. Delides, C. G., and Shephard, I. W., *Radiat. Phys. Chem.*, 10, p. 379 (1977).
189. Squire, D. R., and Turner, D. T., *Macromolecules*, 5, p. 401 (1972).
190. Miller, A. A., *J. Am. Chem. Soc.*, 83, p. 31 (1961).
191. Przbyla, R. L., *Rubber Chem. Technol.*, 47, p. 285 (1974).
192. Mironov, E. I., et al., *Kauch Rezina*, 6, p. 19 (1971).
193. Drobny, J. G., *Technology of Fluoropolymers*, CRC Publishers, Boca Raton, FL, p. 93 (2009).
194. Arcella, V., and Ferro, R., in *Modern Fluoropolymers* (Scheirs, J., Ed.), John Wiley & Sons, Chichester, p. 72 (1997).
195. Böhm, G. G. A., and Tveekrem, J. O., *Rubber Chem. Technol.*, 55, p. 649 (1982).
196. Lyons, B. J., in *Modern Fluoropolymers* (Scheirs, J., Ed.), John Wiley & Sons, Chichester, p. 339 (1997).
197. Kaiser, R. J., et al., *J. Appl. Polym. Sci.*, 27, p. 957 (1982).
198. Vokal, A., et al., *Radioisotopy*, 29, p. 426 (1988), CA 112, 119976u.
199. McGinniss, V. D., in *Encyclopedia of Polymer Science and Engineering*, Vol. 4 (Mark, H. F., and Kroschwitz, J. I., Eds.), John Wiley & Sons, New York, p. 438 (1986).
200. McGinniss, V. D., in *Encyclopedia of Polymer Science and Engineering*, Vol. 4 (Mark, H. F., and Kroschwitz, J. I., Eds.), John Wiley & Sons, New York p. 439 (1986).
201. Logothetis, A. L., U.S. Patent 5,260,351 to E.I. Du Pont de Nemours and Company (November 1993).
202. Marshall, J. B., in *Modern Fluoropolymers* (Scheirs, J., Ed.), John Wiley & Sons Ltd., Chichester, UK, 1997, p. 81.
203. Caporiccio, G., and Mascia, L., U.S. Patent 5,457,158 to Dow Corning Corporation (1995).
204. Drobny, J. G., *Handbook of Thermoplastic Elastomers*, William Andrew Publishing, Norwich, NY, p. 150 (2007).

205. Spenadel, L., Grosser, J. H., and Dwyer, S. M., U.S. Patent 4,843,129 to Exxon Research and Engineering Company (June 1989).
206. Stine, C. R., et al., U.S. Patent 3,990,479 to Samuel Moore and Company (November 1976).
207. Drobny, J. G., *Rubber World*, Vol. 232, No. 4, July, p. 27 (2005).
208. Mehnert, R., Bogl, K. W., Helle, N., and Schreiber, G. A., in *Ullmann's Encyclopedia of Industrial Chemistry*, Vol. A22 (Elvers, B., Hawkins, S., Russey, W., and Schulz, G., Eds.), VCH Verlagsgesselschaft, Weinheim, p. 479 (1993).
209. Mehnert, R., Pincus, A., Janorsky, I., Stowe, R., and Berejka, A., *UV &EB Curing Technology & Equipment*, John Wiley & Sons, Chichester/SITA Technology Ltd., London, p. 22 (1998).
210. Janke, C. J., et al., U.S. Patent 5,877,229 to Lockheed Martin Energy Systems (1999).
211. Shu, J. S., et al., U.S. Patent 4,657,844 to Texas Instruments (1987).
212. Carroy, A. C., *Conference Proceedings of 19th Technology Days in Radiation Curing Process and Systems* (May 22–28, Le Mans, France) (1998).
213. Heger, A., *Technologie der Strahlenchemie von Polymeren*, Akademie Verlag, Berlin, 1990.
214. *Chemistry and Technology of U.V. and E.B. Formulations for Coatings, Inks and Paints* (Oldring, P. K. D., Ed.), SITA Technology Ltd., London, 1991.
215. *Radiation Curing of Polymers* (Randell, D. R., Ed.), The Royal Society of Chemistry, London (1987).
216. *Radiation Curing of Polymeric Materials* (Hoyle, C. E., and Kinstle, J. F., Eds.), ACS Symposia Series, 417 (1990).
217. Garratt, P. G., *Strahlenhartung*, Curt Vincentz Verlag, Hannover, p. 45 (1996). In German.
218. Chapiro, A., *Radiation Chemistry of Polymeric Systems*, Wiley-Interscience, New York (1962).
219. Wellons, J. D., and Stannett, V. T., *J. Polym. Sci.*, A3, p. 847 (1965).
220. Demint, R. J., et al., *Text. Res. J.*, 32, p. 918 (1962).
221. Dilli, S., and Garrnett, J. L., *J. Appl. Polym. Sci.*, 11, p. 859 (1967).
222. Hebeish, A., and Guthrie, J. T., *The Chemistry and Technology of Cellulosic Copolymers*, Springer Verlag, Berlin (1981).
223. Garnett, J. L., *Radiat. Phys. Chem.*, 14, p. 79 (1979).
224. Dworjanyn, P., and Garnett, J. L., in *Radiation Processing of Polymers* (Singh, A., and Silverman, J., Eds.), Hanser Publishers, Munich, Chapter 6, p. 94 (1992).
225. Kabanov, V. Y., Sidorova, L. P., Spitsyn, V. I., *Eur. Polym. J.*, p. 1153 (1974).
226. Huglin, M. B., and Johnson, B. L., *J. Polym. Sci.*, A-1, 7, p. 1379 (1969).
227. Dworjanyn, P., and Garnett, J. L., in *Radiation Processing of Polymers* (Singh, A., and Silverman, J., Eds.), Hanser Publishers, Munich, Chapter 6, p. 99 (1992).
228 Trommsdorf, E., Kohle, G., and Lagally, P., *Makromol. Chem.*, 1, p. 169 (1948).
229. Dilli, S., et al., *J. Polym. Sci.*, C37, p. 57 (1972).
230. Dilli, S., and Garnett, J. L., *Aust. J. Chem.*, 23, p. 1163 (1970).
231. Ang, C. H., et al., in *Biomedical Polymers, Polymeric Materials and Pharmaceuticals for Biomedical Use* (Goldberg, E. P., and Nakajima, A., Eds.), Academic Press, New York (1960).
232. Dworjanyn, P., and Garnett, J. L., in *Radiation Processing of Polymers* (Singh, A., and Silverman, J., Eds.), Hanser Publishers, Munich, Chapter 6, p. 112 (1992).

233. Stannett, V. T., Silverman, J., and Garnett, J. L., in *Comprehensive Polymer Science* (Allen, G., Ed.), Pergamon Press, New York (1989).
234. Stannett, V. T., *Radiat. Phys. Chem.* 35, p. 82 (1990).
235. Dworjanyn, P., and Garnett, J. L., in *Radiation Processing of Polymers* (Singh, A., and Silverman, J., Eds.), Hanser Publishers, Munich, Chapter 6, p. 114 (1992).
236. Garnett, J. L., and Leeder, J. D., *ACS Symp. Ser*, 49, p. 197 (1977).
237. Bett, S. J., and Garnett, J. L., *Proceedings of Radcure'87 Europe*, SME Technical Paper FC-260, Society of Manufacturing Engineers, Dearborn, MI (1987).
238. Simpson, J. T., *Radiat. Phys. Chem.*, 25, p. 483 (1985).
239. Pacansky, J., and Waltman, R. J., *Prog. Org. Coat.* 18, p. 79 (1990).
240. Lapin, S.C., Modification of Polymer Substrates Using Electron Beam-Induced Copolymerization, *UV+EB Technology*, Vol. 1, No. 1, p. 44 (2015)
241. Goldberg, E. P., et al., U.S. Patent 6,387,379 to University of Florida (2002).
242. Salovey, R., et al., U.S. Patent 6,281,264 to the Orthopedic Hospital, Los Angeles, and University of South California, Los Angeles (2001).
243. Perez, E., Miller, D., and Merrill, E. W., U.S. Patent 5,836,313 to Massachusetts Institute of Technology (1998).
244. Hasegawa, T., and Kondo, T., U.S. Patent 6,127,438 to Asahi Kasei Kogyo Kabushiki Kaisha (2000).
245. Nablo, S. V., U.S. Patent 5,825,037 to Electron Processing Systems, Inc. (1998).
246. Kaetsu, I., in *Radiation Processing of Polymers* (Singh, A., and Silverman, J., Eds.), Carl Hanser Publishers, Munich, Chapter 8 (1992).
247. Wendrinsky, J., *Europe 2001*, Basel, Switzerland, October 8–10 (2001).

7

Coating Methods Used in UV/EB Technology

There are numerous methods of applying liquid systems to different surfaces, and the following sections will discuss some of them. Many of these techniques are used in both UV and EB curing processes. Some are very specific to UV technology, and these will be dealt with separately in connection with that curing method.

7.1 Roll Coating

Roll coating is widely used as an efficient way to coat uniform, generally flat or cylindrical surfaces of rigid or flexible substrates. There are many types of roll coating methods; the two most common are *direct roll coating* and *reverse roll coating*.

7.1.1 Direct Roll Coating

In direct roll coating, the sheets or stock to be coated pass between two rollers, an *applicator roller* and an *impression* (or *backup*) *roller*, which are rotated in opposite directions. Thus, the material being coated is pulled by the rolls. The applicator roll is usually covered with a relatively hard elastomeric compound. The coating is fed to the applicator roll by a smaller *feed* or *doctor* roll that is in turn fed by a *pickup roll*. The pickup roll runs partially immersed in a tray called a *fountain*, containing the coating.[1] As the coated material comes out of the nip between the rolls, the wet layer of the coating is split between the applicator roll and the substrate. The film thickness is controlled by the clearance between the feed roll and the applicator roll. The arrangement for direct roll coating is shown in Figure 7.1.

Several variations of direct roll coating are possible by using different types of applicator rolls. For example, the entire surface of the applicator roll may consist of engraved small recessed cells (see Figure 7.2). The cells are filled with coating; the surface is scraped clean with a *doctor blade*; and the coating is transferred to the substrate being coated. This type of roll is referred to as a *gravure applicator roll*.[2] Besides being employed in precision coating, this type of roll finds use in printing (see Section 9.5.5). The engraved roll can be partly immersed in the coating pan so that the fountain and feed rolls are not needed.

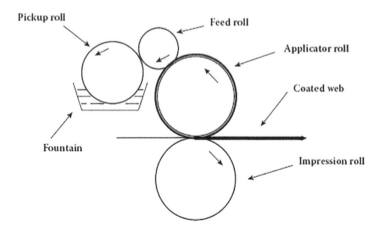

FIGURE 7.1
The arrangement for direct roll coating.

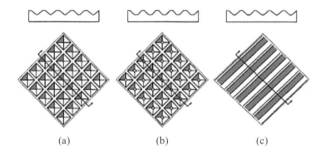

FIGURE 7.2
Different types of cells on the surface of a gravure roll: (a) pyramidal; (b) quadrangular; (c) tri-helical. (Courtesy of Flexographic Technical Association.)

7.1.2 Reverse Roll Coating

In reverse roll coating, the two rolls are rotating in the same direction and the material being coated has to be moved through the nip by means of a drive roll. The coating is applied by wiping rather than film splitting and is picked up by the web moving in the reverse direction. The coat weight can be exactly controlled by the gap between the metering and application rolls. Reverse roll coating is suitable for applying high-viscosity coatings

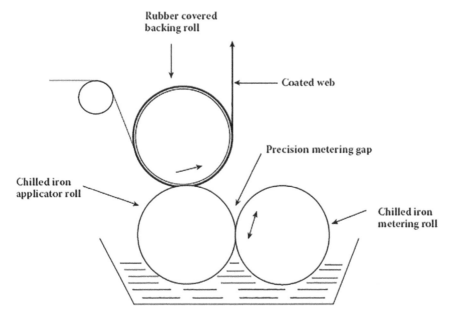

FIGURE 7.3
Reverse roll coating.

onto flexible substrates, and the result is a smoother film than that obtained by direct roll coating.[1,2] The arrangement for reverse roll coating is shown in Figure 7.3. A variety of possible arrangements for roll coatings is shown in Figure 7.4.

7.2 Curtain Coating

Curtain coating is widely used in the coating of flat sheets of substrate, such as wood panels. A coating is pumped through a slot in the *coating head* so that it flows as a continuous *curtain* of liquid. The material to be coated is moved under the curtain by a conveyor belt. The curtain is always wider than the substrate and therefore a recirculating system is necessary for the overflow from the sides.

The film thickness is controlled by the width of the slot and the speed of the substrate being coated; the faster the speed, the thinner is the coated wet film. Because there is no film splitting, the coating is laid down essentially smooth.[1]

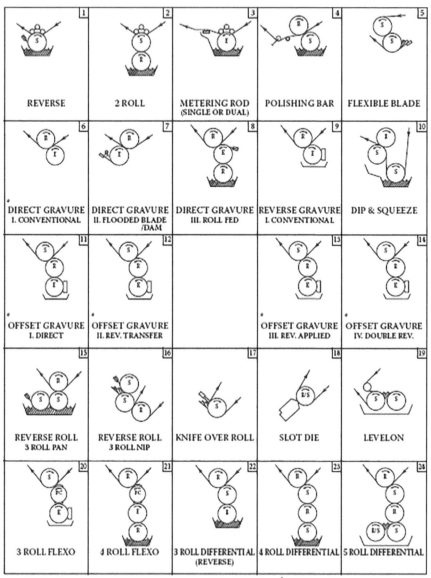

FIGURE 7.4
Different roll coating methods and roll arrangements: S = steel roll; R = rubber-covered roll; E = engraved roll. (Courtesy of Faustel Corporation.)

7.3 Spray Application

Spray application is widely used for applying coatings and paints in industrial and architectural applications. It is particularly well-suited for surfaces with irregular shape, although it can also be used for flat surfaces if other methods, such as roll or curtain coating, would not be feasible. The main disadvantage of spraying is the difficulty of controlling where the paint goes and the inefficiency of application, since only a fraction of the spray particles are actually deposited on the object being sprayed.

There are many different kinds of spraying equipment used for coatings; they all atomize the liquid into droplets. The droplet size depends on the type of spray gun and coating. The variables affecting it are air and liquid pressure, liquid flow, viscosity, and surface tension.

7.3.1 Compressed Air Gun Spraying

The oldest method, still in use today, is employing compressed air guns, in which the stream of liquid is driven through the nozzle orifice by pressures of 1–5 kPa (1.5–7 psi). In general, compressed air guns are less expensive than other spray guns. Atomization can be finer than with other methods. The system is versatile, and virtually any sprayable material can be sprayed with air spray. Improved transfer efficiency is achieved by *high-volume, low-pressure* (HVLP) air guns. These operate at low air pressures, typically 2–7 kPa (3–10 psi), and at a much higher air volume than standard spray guns.[1] Spray guns can be handheld or attached to robots.

7.3.2 Airless Gun Spraying

In airless spray guns, the liquid is forced out of the orifice at high pressure, anywhere from 5 to 35 MPa (725 to 5,075 psi). As the liquid comes out of the orifice, the pressure is released, resulting in cavitation that leads to atomization. The atomization is controlled by the viscosity (higher viscosity, larger particles), pressure (higher pressure, smaller particle size), and surface tension (lower surface tension, smaller particle size). Air-assisted airless spray guns are also available; the atomization is airless, but there are external jets to help shape the fan pattern.[3] Both handheld and robot airless guns are available.

7.4 Dip Coating

The object to be coated is dipped into the tank full of coating and pulled out, and the excess coating drains back into the dip tank. To minimize the thickness differential, the rate of withdrawal is controlled.

7.5 Flow Coating

Objects to be flow coated are coated on a conveyor through an enclosure in which streams of coating are squirted on them from all sides.[3] The excess material runs off and is recirculated through the system. There is still some thickness gradation but much smaller than in dipping.

7.6 Spin Coating

Spin coating utilizes centrifugal forces created by a spinning substrate to spread a liquid evenly over its surface.[4] Current applications are in photoresist technology for the microelectronic industry and in the manufacture of protective overcoats and adhesives for the optical storage industry (compact disks and DVDs).[4]

7.7 Rod Coating

Wire-wound rod (Meyer rod) is suitable for coating flat, flexible surfaces with low-viscosity liquids. Its function is to remove excess liquid from the moving substrate, thereby leveling the surface of the substrate. This surface shaping is often used to achieve high-gloss coating. Usually, the rod rotates in order to avoid collection of dirt particles on it. The wet thickness of the coating depends on the diameter of the wound wire, fluid rheology, web speed, and rotation of the rod.[5] Similarly, smooth bars, blades, or air knives are used for the same purpose. If coatings applied by these

TABLE 7.1

Coating Methods, Characteristics, and Corresponding Coating Thickness

Coating Method		Coating Viscosity, Pa.s	Coating Speed, m/min (ft/min)	Coating Thickness, μm
Roll coating	Forward	0.2–1	300 (985)	10–200
	Reverse	0.1–50	150 (495)	4–400
Gravure		0.001–5	700 (2300)	1–25
Rod (incl. wire wound)		0.02–1	250 (820)	5–50
Cast coating		0.1–50	150 (490)	5–50

methods have insufficient surface smoothness, casting against a polished chrome-plated drum or a quartz cylinder is a relatively simple method to achieve coating surface with nearly optical quality. The topology of the casting cylinder surface is reproduced by the coating brought to the contact with the cylinder. Within a certain angle of rotation the coating is irradiated by a UV source, and the coated web is removed from the drum.[5]

Some coating methods and their characteristics are summarized in Table 7.1.[5,6] The numbers are only rough guidelines.

7.8 Vacuum Coating

In vacuum coating, the substrate is fed through an entry template into the application chamber and leaves through an exit template. A vacuum pump or pumps generate vacuum in the system. The negative pressure causes air to be pulled into the application chamber through the entry and exit templates.

This air causes the coating material to be brought into contact with the substrate.[7] A diagram of vacuum coating system is shown in Figure 7.5.

The excess coating is removed and is pulled back into the baffle tower, and air and coating are separated. Air is pulled through the system toward the vacuum pump. The coating returns to the reservoir for recycling. The coating weight can be controlled by the regulation of the vacuum: the higher the vacuum inside the system, the less coating is applied to the

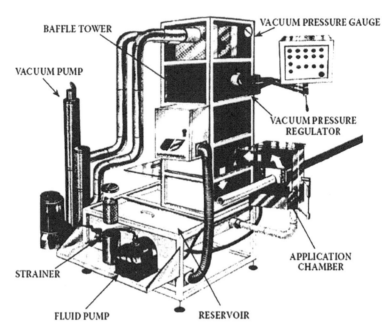

FIGURE 7.5
Diagram of vacuum coating system. (By permission from RadTech International, N.A.)

substrate. Vacuum coating is best suited to applying coatings to moldings and round (pole-type) products. Its major advantage is that all four sides of the molding are coated in one pass.

References

1. Wicks, Z., in *Organic Coatings: Science and Technology*, Chapter 1 (Wicks, Z., Ed.), Wiley-Interscience, New York (1994).
2. Pasquale III, J. A., *Converting*, 19, No. 12, December, Special Report, p. 62 (2001).
3. Easton, M. E., *J. Oil Colour Chem. Assoc.*, 66, p. 366 (1983).
4. Taylor, J. F., *Spin Coating: An Overview and Guide*, www.radtech.org/Applications (September 2000).
5. Mehnert, R., Pincus, A., Janorsky, I., Stowe, R., and Berejka, A., *UV&EB Curing Technology & Equipment*, Vol. I, John Wiley & Sons Ltd., Chichester/ SITA Technology Ltd., London, p. 184 (1998).
6. Gutoff, E.B., in *Modern Coating and Drying Technology*, Chapter 2 (Cohen, E. D., and Gutoff, E. B., Eds.), VCH, Weinheim (1992).
7. Whittle, K., *RadTech Report*, 18, No. 4, July/August, p. 18 (2004).

8

Applications of UV Radiation

In previous chapters, it was shown that both ultraviolet radiation (photons) and electron beam energy (high-energy electrons) have the capability to convert certain liquid systems into solids. Although there are some areas of industrial applications, where both curing methods work, the choice of the equipment and process will depend on several factors, such as capital cost, production volume, and variety of products. Electron beam equipment usually represents a major capital investment and favors large-volume production of single or not more than a few products. On the other hand, it is capable of penetrating thicker layers, both clear and opaque, and generally works with simpler chemistry. Recent developments in both technologies, such as improved UV equipment and photoinitiators on the one hand, and low-voltage, much less expensive EB processors on the other, have further increased the number of choices. Thus, the final selection can be made only after a thorough analysis.

In this chapter, we will focus on applications where at the current state of technology UV radiation offers pronounced advantage or is the only known method for that application. Chapter 9, dealing with applications using EB, is written in the same fashion.

In contrast to currently available electron beam equipment, UV curing hardware is available in a multitude of different sizes, designs, power, in-line, off-line, in combination with large coating or printing machines, together with alternative drying equipment, as well as stand-alone units. The selection of the type of UV curing equipment is then done with regard to the process, including the substrate, the coating, paint, adhesive, or ink to be used, and the configuration of the part to be irradiated. Substrates to be irradiated can be two-dimensional, cylindrical, and three-dimensional, or the application can require only the irradiation of small areas (spot cure). The UV curing system can also be part of robotics applications.

It should be noted that at this writing the UV light-emitting diodes (UV-LEDs) are being used as an alternative and a replacement to traditional mercury arc lamps. They offer a multitude advantages over the mercury arc systems (see Section 3.1.7), and it is quite possible that they might replace them entirely. In fact, most of the applications discussed in this chapter can use UV-LEDs without any major changes. For that reason, we are not writing separate sections for the use of UV-LED equipment.

The main technical and design issues to be resolved are[1]:

- Cure efficiency
- Heat management
- System reliability
- Suitable integration into the production line

The curing equipment will operate at the sufficient efficiency if a minimum UV dose (mJ/cm^2) or a minimum exposure time can be applied to obtain the desired degree of cure. This requires that the reactive binders and photoinitiator system closely match the emission spectrum of the UV source.

As discussed in Section 3.1, part of the electrical energy is converted into IR radiation that may be damaging to the substrate being processed. Air or water cooling and reflector design are the main factors involved. Heat management also means instantaneous stop of irradiation of a substrate that is not moving. In most cases, a properly designed shutter prevents overheating the standing substrate. When the shutter closes, the lamp power is automatically reduced to standby mode. This of course does not apply to the case in which UV-LED equipment is used.

The system reliability not only means minimum downtime but also implies a compliance with the quality of the curing process, and it is the manufacturer's responsibility to design it to meet these criteria.[2]

When integrating the UV curing system into the production line, attention must be paid to the following[3]:

- Master-slave operation of machine and UV curing unit in all states of operation (start, standby, and production)
- Status monitoring of electric power, shutter and reflector operation, temperature, and cooling conditions
- Easy maintenance and servicing
- Minimum effect of the UV unit on machine parts (no excessive heating or ozone-induced corrosion)

Uniformity of the supplied energy from the radiation source is crucial to produce a uniform finished coat. There are essentially three possible arrangements:

- A *single array* may be a single lamp sufficiently long to cover the width of the web being processed, or several shorter sources covering the width. Clearly, such an arrangement is used for flat or slightly curved webs or very simple shapes. This setup is used widely for printing or wood finishing.

- A *multiple array* usually consists of 10–12 lamps and adjustable carriages. It is suitable for irradiating parts with complex shapes or many different shapes of parts. The arrangement allows focusing the light on all coated surfaces.
- The *robotic array* is best suited for complex geometries. It uses the same type of robot that applied the coating, and the system has a single source of radiation for each robot. An example of using robotics for UV cure is shown in Figure 8.1.

UV technology is becoming more dispersed over more markets due to its acceptance in markets that did not use it a decade ago, such as metal coatings and exterior applications. The wood market is still expanding mainly due to environmental concerns (primarily emissions) and improving efficiency.

On the other hand, the automotive original equipment manufacturer (OEM) sector is the slowest market due to the extensive testing that has to be done on coatings. Another factor here is the recent slowdown of the economy in general and of the automotive industry in particular.

The refinishing side of the automotive market is growing considerably faster because the applications have been developing over the past decade and represent numerous advantages, such as efficiency and improved environment, due to drastic reduction of the use of solvents and due to improved practices.

FIGURE 8.1
3D UV curing using robotics. (Courtesy of IST Metz.)

8.1 UV Curing of Coatings and Paints

UV curing of coatings and paints represents the largest volume for the application of that process. There are many types of coatings and paints with a variety of formulations, viscosities, and physical properties, which can be either functional or decorative. These paints and coatings are applied onto a large variety of substrates with a diversity of properties and shapes.

8.1.1 Functional and Decorative Coatings

8.1.1.1 Coatings on Flat, Rigid Substrates

UV curable coatings are applied to a variety of flat, rigid substrates, such as particleboard, medium- and high-density fiberboard, wood veneers, polycarbonate, poly(methylmethacrylate), paper, and metal sheets and foils. An example of a formulation for UV curable coatings is shown in Table 8.1.

UV curing units are an essential part of the coating line, consisting typically of the following main components[4,5]:

1. Infeed
2. Substrate preparation
3. First coating station
4. First UV curing unit
5. Second coating station
6. Second UV curing unit
7. Last coating station
8. Last UV curing unit
9. Outfeed

TABLE 8.1

Example of a UV Coating Formulation

Ingredient	Typical Amount, Weight %	Function
Prepolymer (oligomer)	25–95	Film forming, affects film properties
Monomer (reactive thinner)	0–6	Viscosity adjustment, contributes to film forming
Fillers and/or pigments	0–60	Cost saving, coloring, improve sanding
Photoinitiator	1–5	Initiates UV-induced polymerization
Miscellaneous additives, defoamers, matting agents	0–3	Affect processing behavior and appearance of finished film

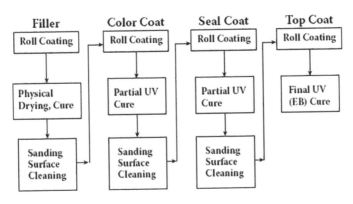

FIGURE 8.2
Example of UV curing of wood coatings. (Adapted from Mehnert, R. et al., *UV&EB Curing Technology & Equipment*, Vol. I, John Wiley & Sons, Chichester/SITA Technology Ltd., London, p. 173, 1998.)

As a general procedure, the first step is preparation of the substrate, which may include cleaning, sizing, and applying thin overlay by lamination to cover surface imperfections. Metal surfaces often need a primer coat to ensure a good bond of the first coat to the metal. Then the first coat is applied to the pretreated substrate. Subsequently, a second coat, often a pigmented color, is applied if desired. The next coat is a sealer, followed by a topcoat. An example of a UV curing process in wood coating is shown in Figure 8.2.

For flat surfaces, the application method depends on the coating weight. Roll coating and curtain coating are the most efficient and economical methods to apply UV coatings to flat surfaces. The maximum coating weight that can be applied by a roll is 40 g/m²; the limit for curtain coatings is 80 g/m.[2,6]

Normally, the last UV unit should ensure an adequately thorough cure. This may be a problem for thick layers, particularly if they are pigmented. In such cases, electron beam irradiation would be the preferred curing method.[6]

8.1.1.2 UV Curing of Coatings on Flexible Substrates

Coatings for flexible substrates are mostly reactive systems with no solvents. Since they are usually highly viscous, they tend to produce a rough surface.

Therefore, the selection of the proper application method is crucial. The selection is determined by the following factors[7]:

- Coat weight (layer thickness)
- Coat viscosity and viscoelasticity
- Coat weight accuracy required
- Coating speed
- Substrate to be coated

FIGURE 8.3
Water-cooled UV curing system for flexible packaging. (Courtesy of Prime UV-IR Systems.)

Roll coating is the preferred method for most coatings applied to flexible substrates. Gravure roll coating is the most precise method, yet allowing coating speeds up to 700 m/min (2,135 ft/min).[8] Other roll coating methods, such as forward and reverse coating, are selected according to the task and it depends on the viscosity of the coating and coating thickness (see Chapter 6). For higher precision and lower coating weights, multiple roll (typically four or five) systems are often employed.[9] An example of a modern coating line with multiple curing lamps is shown in Figure 8.3.

8.1.2 UV Curing of Lacquers, Varnishes, and Paints

The coatings include clear overprint varnishes, finish coatings for PVC, wax-less flooring, finish coatings for paper and film applied as laminates in wood decoration, and transparent functional coatings with special characteristics, such as high abrasion resistance, barrier properties, conductivity, and chemical resistance. Coat weights of clear varnishes and lacquers may vary from 2 to 5 g/m^2 for typical overprint applications and from 10 to 100 g/m^2 for finish and functional coatings. Because of the relatively low coating weights, gravure and reverse roll coating methods are used in most cases.[10]

8.1.3 Three-Dimensional Curing

Because of their frequently complicated shapes, three-dimensional (3D) objects require different coating techniques, such as spray, dip, and flow and spin coating. The lamps in curing units for 3D objects are stationary, and the parts pass in front of them. They usually rotate two to three times as they pass through the irradiation zone. The irradiation zone must have a shield to eliminate direct and first scattered light. Because of the complexity of coated parts and high demands on quality, UV lamps are frequently mounted on robotics (shown previously in Figure 8.1).

8.1.4 UV Curing of Coatings and Inks on Cylindrical-Shaped Parts

This method is used for curing coatings and inks on plastic cups, tubs, tubes, or metal cans.[11] The parts are placed on mandrels, which are attached to a rotating device. This device moves them through the individual stations: feed, pretreat (most frequently corona or flame, for plastics), printing, curing, and takeoff. The printing is done by dry offset (see Section 8.5.3).

Compact and high-powered UV lamps are used to allow curing of the rotating cup directly on the mandrel. Curing times are usually in the range of 70–100 ms.[12]

8.1.5 UV Matting of Coatings

The impression of a matte film surface is created by the diffuse reflection of light from a micro-rough surface. Such a structure is obtained by incorporating a matting agent (e.g., synthetic amorphous silica) or by changing the surface topology through shrinkage during polymerization.

To achieve matte surface by UV curing, traditionally, the so-called *dual-cure process* is used.[13] In the first step, the coated film is irradiated by a low dose, and the matting agent (most frequently synthetic amorphous silica) in the coating can migrate to the surface of the film. The film is then cured completely in the second step. With acrylate-based coatings, the method is modified by using metal halide lamps for the first irradiation step, whereas the final surface cure is accomplished by mercury arc lamps. Another method is based on cure inhibition by oxygen. In the first step, a matted surface is created by surface cure inhibition and the surface structure is fixed by the second cure (UV or EB).

A relatively new method uses 172 or 222 nm excimer radiation in nitrogen atmosphere.[14] The irradiation causes a surface cure of the acrylate coating similar to skim milk in appearance. In the second step, the coating is fully cured by UV or EB.

Each method and application requires specific formulation with specific properties, such as solids content, density, viscosity, color, finished film thickness, scratch resistance, gloss, etc.

8.2 UV Curing of Adhesives

Radiation curable adhesives represent a small but growing segment of the overall radiation curing market because they are an attractive alternative to solvent-based, water-based, and hot melt adhesives. Structural adhesives, pressure-sensitive adhesives (PSAs), laminating adhesives, and transfer metallization adhesives can be radiation-cured. Commercial adhesive products made on UV/EB equipment include pressure-sensitive tapes and labels, laminated foils and films, flocked materials for automotive and shoe applications, structural bonding adhesives, and abrasive bonding systems. The technological and cost advantages of these adhesives are that they are single-component materials that can be dispensed with automatic dispensing equipment, have long open time and low-energy requirements, exhibit fast cure rates, have the ability to precision bond, and exhibit low heat input to temperature-sensitive substrates.[15]

8.2.1 Energy-Curable Laminating Adhesives

Laminating adhesives generally bond clear laminate plastic films. As long as the films are clear, the adhesive can be cured by UV energy; if they are opaque, electron beam radiation is a better choice.

PSA-type laminating adhesives can be cured prior to the lamination. The adhesive film is tacky as a typical PSA and produces good bonds. Another method involves the lamination first and the cure after that. In this case, the bond strength depends mainly on the adhesion of the adhesive to the substrate; the cured adhesive film is strong, and consequently, the cohesion of the laminate is high. In many cases, the bond is so strong that one or both substrates fail.[16]

Examples of laminating adhesives based on urethane acrylates are shown in Tables 8.2 and 8.3. The respective resulting bond strength values from varied substrates are shown. Specimens were prepared in the following fashion:

TABLE 8.2

Formulation for UV Curable Lamination Adhesive

Component	% wt.	Remark
Inert resin in TPGDA[a]	60.0	Base resin (elastomer)
Dipropylene glycol diacrylate (DPGDA)	34.0	Difunctional reactive diluent
Methacrylated phosphate ester	3.0	Adhesion promoter
Aromatic ketone	3.0	Photoinitiator
Total	100.0	

Test Results:

Viscosity, mPa.s:	450
Reactivity, mJ/cm^2:	100
Adhesion, T-peel, N/25 mm	
OPP/OPP:	SF
Metallized OPP/PE:	SF
PE/PE:	3.78
PE/Paper:	SF

Abbreviation: SF = Substrate failure.

Specimen preparation: Film thickness 0.3 mil (7.7 μm); the substrate was corona treated to 42 dynes/cm; adhesive cured through the laminate film using a 300 wpi (120 W/cm) mercury lamp.
[a] Tripropylene glycol diacrylate, reactive thinner.

TABLE 8.3

Formulation for PSA-Type UV Curable Laminating Adhesive

Component	% wt.	Remark
Monofunctional urethane acrylate in EHA[a]	26.1	Base resin (elastomer)
Modified saturated polyester resin in EHA[a]	35.1	Tackifier
2-(2-ethoxyethoxy)ethyl acetate	23.6	Inert diluent
Isobornylacrylate monomer (IBOA)	5.6	Diluent monomer
Aliphatic urethane acrylate	5.6	Reactive diluent
Aromatic ketone	4.0	Photoinitiator
Total	100.0	

Test Results:
Viscosity, mPa.s: 560
Reactivity, mJ/cm^2:   500
Adhesion, T-Peel, N/2.5 mm

OPP/PE:	19.34
PE/PE:	10.45
PET/PE:	7.78
Metallized OPP/PE:	6.67
PE/Paper:	SF

Abbreviation: SF = Substrate failure.

Specimen preparation: Film thickness 0.3 mil (7.7 µm); the substrate was corona treated to 42dynes/cm; adhesive cured through the laminate film using a 300 wpi (120 W/cm) mercury lamp.

[a] 2-ethylhexyl acrylate monomer.

8.2.2 Energy-Curable Pressure-Sensitive Adhesives

UV curable PSAs are based on two types of UV curable materials:

1. Materials that already are pressure-sensitive adhesives but can be modified by radiation (UV or EB) to introduce or increase cross-linking, which increases their adhesive properties, shear value, peel strength, service temperature range, and solvent resistance.

2. Materials, such as acrylates with tackifiers, acrylate polyesters, and polyurethanes, that do not have properties as pressure-sensitive adhesives but after cross-linking have desired pressure-sensitive properties.

In the past, electron beam radiation was applied to produce PSAs exclusively; however, recent improvements in UV curing technology (precise UV dose control, suitable photoinitiators) permit UV to be used to produce pressure-sensitive adhesives. PSA formulations can vary in consistency from low-viscosity liquids up to solids melting at 80°C (176°F). Therefore, applications may vary from screen printing to roll coating to melt extrusion. Coat weights for most PSA materials vary from 1 to 10 g/m^2.

As pointed out above, radiation or energy-curable adhesives can be essentially hot melts or liquids at room temperature. For this technology, a liquid curable PSA formulation is composed of four essential components:

1. Base resin (elastomer)
2. Tackifier
3. Diluent
4. Photoinitiator

An example of the *elastomeric component* is a combination of multifunctional and difunctional aliphatic urethane oligomers. Suitable oligomers have relatively high molecular weights and glass transition temperatures (T_g), which enable the adhesive to have elastic properties at room temperature. Its deformability under light pressure allows it to conform to and wet out a substrate. Upon removal of the adhesive from the substrate, its elasticity allows it to extend greatly before separating, giving it a good tack and peeling properties.

The *tackifier* is most frequently a saturated polyester resin. It generally has a much lower molecular weight and higher T_g. The higher T_g of the tackifying resin brings the overall adhesive T_g to a value necessary to achieve pressure-sensitive properties. Typically, the T_g values of PSAs range from –25°C to +5°C. In general, the higher the overall T_g of an adhesive, the greater the cohesive strength and high-temperature shear results, and the lower the tack and deformability properties.

The reactive monomer, besides being a *diluent*, plays a role similar to that of tackifying resin, namely, affecting the deformability of the adhesive.

The *photoinitiator* (single or in a combination) is vital to the manner in which an adhesive cures. The desired property of the PSA is a balance of good tack and good adhesion. If the state of cure is too high, the tack will be too low, and if the state of cure is too low, the adhesive will stick to the substrate but will be difficult or impossible to remove.

The methods to evaluate PSAs include the *rolling ball test* (ASTM D3121, PSTC-6, BS EN 1721), *loop tack test* (ASTM D6195, FINAT Test Method 9, BS EN 1719), and *quick stick test* (PSTC-5).[16]

Pressure-sensitive adhesives are used in a great variety of applications, most commonly for adhesive tapes. In that case, they have to be tested by a *static shear test* or *dynamic shear test*.[16] The difference between these two methods is that in a static shear test a standard force is being applied to the test specimen and the adhesive failure is reported as the time it takes for failure to occur. The dynamic shear test involves a force being applied to the PSA tape at a specific rate of speed (typically 0.25 mm or 0.1 in. per minute). The value reported is as the peak force per unit area ($lb/in.^2$, also abbreviated as psi, and in SI units MPa) required to cause adhesive failure. The standards for adhesion shear tests are ASTM D3654, ISO EN 1943, and PSTC-107.[17,18]

Another type of shear test involves testing under heated conditions. The shear adhesion failure test (SAFT) (FINAT FTM 8, PSTC-7) is a static shear test administered to a PSA sample under increasing temperature at a rate typically of 1°C or 2°C per minute. The recorded SAFT value is the temperature at which the adhesive bond fails.

The adhesion peel test (ASTM D3330, ISO EN 1939, PSTC-101) determines the force required to remove the respective adhesive from a specific substrate. Three components are involved in this process: the extension of the adhesive itself, the deformation of the PSA backing substrate during the removal process from the substrate surface and the force to remove the adhesive from the substrate. The adhesive tapes are generally removed under a 90° (T-peel test, ASTM D1976) or 180° angle at a specific rate. The result is shown as the average peeling force per unit width, i.e., lb/in. (sometimes shown as pli, and in SI units as kN/m).

An example of the *high-tack UV curable* PSA formulation is shown in Table 8.4. The low T_g and cross-linking combination of this PSA offers very high tack performance but limited cohesion, which is evident in the low shear value and negligible heat resistance. A typical *well-balanced UV curable* PSA formulation is shown in Table 8.5. This formulation offers a well-balanced PSA performance with exceptional adhesion and shear properties. This is accomplished by the use of inert tackifying resin to allow the oligomeric component to maximize its elastic characteristics and to decrease the cross-link density, which increases the tack and peel properties while still maintaining very good shear and heat resistance.[19]

TABLE 8.4

High Tack UV Curable PSA Formulation

Component	% wt.	Remark
Monofunctional urethane acrylate in EHA[a]	17.0	Base resin (elastomer)
Inert resin in aliphatic urethane acrylate monomer	55.0	Tackifier
2-(2-ethoxyethoxy)ethyl acetate	22.0	Diluent
Aromatic ketone	6.0	Photoinitiator
Total	100.0	

Test Results:

Viscosity, mPa.s:	1,500
Loop tack, lb/in^2 (MPa):	3.49 (2.4×10^{-2})
Peel strength, pli (kN/m):	2.54 (0.43)
Static shear strength at RT, h:	1.25
Static shear strength at 80°C, h:	0.00

[a] 2-ethylhexyl acrylate monomer

TABLE 8.5

Well-Balanced UV Curable PSA Formulation

Component	% wt.	Remark
Monofunctional urethane acrylate in EHA[a]	40.0	Base resin (elastomer)
Difunctional aliphatic urethane acrylate in dilutant	10.0	Base resin (elastomer)
Inert resin	25.0	Tackifier
Difunctional oligoamine resin	3.0	Reactive resin
Isodecyl acrylate	18.0	Increases cross-linking
Aromatic ketone	4.0	Photoinitiator
Total	100.0	

Test Results:

Viscosity, mPa.s: 5,900
Loop tack, lb/in^2 (MPa):   1.70 (1.2×10^{-2})
Peel strength, pli (kN/m):   6.40 (1.1)
Static shear strength at RT, h: Indefinite
Static shear strength at 80°C: Indefinite
[a] 2-ethytfa acrylate monomer.

8.2.3 Energy-Curable Assembly Adhesives

UV curable adhesives have been used in electronics and electrical manufacturing operations for many years. There are many uses, such as bonding of coils, bonding of loudspeaker membranes, bonding and sealing of wires in ducts, bonding of liquid crystal displays, and bonding of membrane switches.[20]

Another well-established use of UV/visible light-curing adhesives is glass, a plastics bonding in the automotive industry. Typical applications are lamination of safety glass, fastening of rear-view mirrors to windshields, and assembly of headlights.[20]

In the medical field, UV curable adhesives are used in the assembly of medical devices such as syringes, valves, manifolds for filtering equipment, and arteriographs. In these applications, many dissimilar substrates such as stainless steel, aluminum, glass, polycarbonate, polymethyl methacrylate (PMMA), PVC and other thermoplastics are bonded.[21]

Light-curing cyanoacrylate adhesives are suitable for high-speed assembly processes. These one-part, solvent-free systems offer rapid cure of exposed and confined adhesive on bond joints without requiring the use of heat or racking of parts. The bonds exhibit good environmental and thermal resistance, high strength, and a wide range of physical properties. Since they also cure by absorption of moisture, this dual-curing system reduces cure problems in shadow areas. They cure via an anionic mechanism that is not inhibited by oxygen. With special primers, cyanoacrylates form good adhesive bonds with polyolefins and fluoropolymers. The limitations of light-curing cyanoacrylates are blooming/frosting; they may cause stress cracking of some plastics; and they are limited to cure through a gap of about 10 mils (0.25 mm).[22]

8.3 UV-Cured Silicone Release Coatings

Silicone release liners are used as a non-adhering surface to which adhesive materials such as, for example, pressure-sensitive adhesives, can be laminated. UV curable release coatings are based on poly(dimethylsiloxane) oligomers functionalized with acrylate or epoxy groups and are mostly coated onto papers or thin films.

The coat weights are critical for the release properties of the liner. They are typically in the range from 0.7 to 1.0 g/m^2 and are applied most frequently by multiroll reverse coating systems or by offset gravure rolls.

The chemistry involved depends on the oligomer used: silicone acrylates cure by radical mechanism; epoxy-silicones require cationic curing systems. Silicone acrylates are exceptionally sensitive to oxygen and require nitrogen inertization in the curing zone. Epoxy silicones are not oxygen sensitive and therefore the cure can be done in air. However, for high-speed applications, the humidity in the reaction zone should be kept low to ensure stable curing conditions.[23]

Since temperature-sensitive substrates, such as paper and thin thermoplastic films, and high line speeds (up to 600 m/min or 1,970 ft/min) are used, irradiator heat management and power management are of great significance. The curing systems can consist of medium-pressure mercury lamps or electrodeless medium-pressure lamps. If needed, the irradiators are placed above a cooled drum carrying the substrate. Dielectric barrier-discharge-driven excimer lamps can be also employed as a completely cold source of radiation for exceptionally sensitive substrates (Girling, C., private communication). A nitrogen inerted UV curing chamber used for silicone release coatings and specialty coatings is shown in Figure 8.4.

FIGURE 8.4
Nitrogen inerted curing chambers for silicone release coatings and specialty UV inks.

8.4 Spot Curing

Spot curing is used in curing adhesives and dental compositions. Dental curing systems use 35–75 W tungsten filament quartz halogen tubes. The wavelength range used in such a system is 400–500 nm, with a typical irradiance of 500 mW/cm². Adhesives are typically cured with a short 100 W arc mercury lamp coupled to a UV transmitting waveguide (Girling, C., private communication). There are other lamp systems used for different applications. An example of a spot curing lamp is shown in Figure 8.5, and a small-area vertical curing system operating with a 200 W mercury-xenon lamp at 365 nm is shown in Figure 8.6. These systems are used in the electronic industry (e.g., for assembly of printer heads) and in medical applications (e.g., attaching dispensing needle to the syringe barrel).[24] A flood curing lamp used for somewhat larger areas is shown in Figure 8.7. LED curing equipment is shown in Figures 8.8 and 8.9.

FIGURE 8.5
Spot curing lamp. (Courtesy of Dymax Light Curing Systems.)

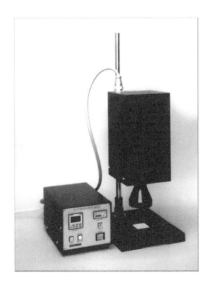

FIGURE 8.6
Mercury-xenon lamp for small area curing. (Courtesy of Hologenix Company.)

FIGURE 8.7
Flood curing lamp. (Courtesy of Dymax Light Curing Systems.)

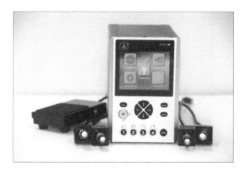

FIGURE 8.8
LED curing lamp. (Courtesy of American Ultraviolet Company.)

FIGURE 8.9
Floor cure LED equipment. (Courtesy of American Ultraviolet Company.)

8.5 UV Curing in Printing and Graphic Arts

The objective of printing is to create a visibly identifiable image, consistently, for a large number of impressions. In principle, this can be done with a printing plate, and the various printing methods are named after the nature of the printing plate. Many techniques have been developed for this purpose.

Flexography, lithography, and gravure are the main printing techniques, which account for the vast majority of printing applications. Each of these methods has a number of variations. For example, the viscosity of printing inks has to be matched to the printing method. Typical values of viscosities of printing inks are shown in Table 8.6.

Graphic arts are a well-established application of UV curing worldwide. Essentially, the processes used in graphic arts include the generation of

TABLE 8.6

Printing Ink Viscosity Ranges

Printing Method	Viscosity @ 25 °C, mPa.s
Gravure	30–200
Flexography	50–500
Newsprint ink	200–1000
Screen printing	1000–50000
Letterpress	1000–50000
Lithography	10000–50000

images to be reproduced onto the printing plate, silkscreen, etc., and the use of radiation curable inks and overprint varnishes. Many imaging processes rely on the exposure of materials to radiation to bring about the change in solubility in a solvent system (organic or aqueous), thus enabling exposed and unexposed areas to be differentiated. The differentiation between exposed and unexposed areas can lead to selective delamination, softening, tackiness (which may affect the adherence of toner powders), or a change in refractive index, which leads to holographic effects.

8.5.1 Screen Printing

The screen printing process uses a rubber squeegee to force ink through a tightly stretched, finely woven mesh onto various substrates. The basic concept has been refined to the point where large semi and fully automatic lines are relatively common. Typical products are display printing, industrial and container printing, and printed circuit production.[24]

Traditionally, evaporation-drying and solvent-based inks have been used. These have been replaced during the last decade by UV curable inks, which exhibit the following advantages[25]:

- UV curable inks can be applied through a much tighter mesh, which produces a lower film weight, improves resolution of fine detail, and lowers ink costs.
- They are more versatile than the traditional inks, and fewer grades of inks are needed for sufficient adhesion to a variety of substrates.
- An almost instantaneous cure reduces seeping into substrate.
- High web speed under UV lamp greatly reduces distortion of substrate.
- There is no drying on screens; thus, screen cleaning is all but eliminated.
- Absence of solvents can eliminate the need for costly solvent recovery systems.

- UV dryers for display printers take up only a fraction of the space occupied by traditional hot-air-jet dryers.
- They are more abrasion and solvent resistant than the traditional inks.
- These inks are friendly to the environment, with practically no health hazards.

Disadvantages cited are cost of raw materials and cost of installation of the UV curing equipment.[25]

Screen printing of UV inks is being done over paper, metal, films, foils, plastics, and PVC. Radiation curable materials are being used as vehicles for ceramic inks that are screened onto automotive windshields, cured, and then fired in an oven to burn off the organic binder and fuse the ceramic into the glass windshield.[26]

8.5.2 Flexography

Flexography is a mechanical printing process that uses liquid ink and a fairly soft relief image printing plate made of rubber, or more commonly photopolymer and pressure, to create an image. Completed printing plates are mounted with adhesive onto a metal cylinder that rotates against the substrate during printing. A typical flexographic printing station is shown in Figure 8.10. A chambered doctor blade ink fountain applies ink to the engraved *transfer*, or *anilox, roll*. The engravings on the anilox roll meter the correct amount of ink, depending on the engraving geometry and depth. The ink is transferred to the raised surfaces of the printing plate attached to

FIGURE 8.10
Flexographic printing deck. (Courtesy of Foundation of Flexographic Technical Association.)

the plate cylinder. The substrate is passed between the plate cylinder and the impression cylinder to achieve ink transfer.

The process is used mainly for package printing. All flexographically printed packaging substrates are web-fed with the exception of corrugated board. Unlike other major mechanical printing processes (namely, lithography, gravure, and letterpress), flexography is in a phase of rapid development and change. Flexography is currently the major package printing process, and it is very unlikely that this position will change over the next few years, even in the light of digital printing technology.

The greatest strengths of UV flexography are its versatility and cost-effectiveness. Since UV inks do not dry until they pass under the UV lamp, the ink stays completely open. Unsightly print defects such as dot bridging caused by dried ink on the printing plate are all but eliminated. In summary, the main advantages of UV flexography over traditional technology are[27]:

- Improved repeatability for printed spot colors
- High-definition process printing: better dot structure, lower dot gains
- Efficient job setup time

The major use for UV flexo printing is a wide range of packaging products, including:

- Corrugated cases
- Folding cartons
- Liquid packaging containers
- Flexible packaging
- Multiwall sacks
- Paper bags
- Labels
- Paper cups
- Containers and closures.

UV inks have been readily and successfully adopted by flexographic label printers; many of them had several years of experience with letterpress UV systems. Since in flexography the common thickness of the printed ink is typically 2–4 µm, it is often necessary to cure individual layers of inks by UV in multicolor printing. With radical UV curable flexographic printing inks, it is possible to achieve printing speeds of 250 m/min (750 ft/min) or even more.[28] A modern flexographic folding carton press with three UV processors is shown in Figure 8.11. An air-cooled in-line UV curing is shown in Figure 8.12, and a system of UV curing of coatings on a wide web high-speed flexo press is shown in Figure 8.13.

FIGURE 8.11
Modern flexographic folding carton press with three UV processors. (Courtesy of Prime UV-IR.)

FIGURE 8.12
Air-cooled in-line UV curing system on a narrow web flexo press. (Courtesy of Prime UV-IR.)

8.5.3 Letterpress and Offset Letterpress (Dry Offset)

The letterpress uses hard plastic or sometimes metal plates[29] and works with heavy pastes that have to be metered by a complicated series of rollers. Cylindrical objects such as two-piece metal cans have no point against which to register printing stations. This can be resolved by using

FIGURE 8.13
UV curing of coatings on a wide web high-speed flexo press. (Courtesy of Prime UV-IR.)

a process known as *offset letterpress* or *dry offset*. In dry offset, the inked images from the letterpress printing plates are transferred onto an inter-mediate resilient rubber blanket roll, where all colors can be assembled in complete register. The blanket roll is then rolled against the round object, and all the colors that make up the image are transferred at one time and in complete register. Offset letterpress is shown schematically in Figure 8.14.

Dry offset is used to decorate drawn and ironed metal cans, round plastic tubs, and plastic or metal collapsible tubes.[29]

8.5.4 Lithography

Lithography is a planographic process, meaning that the printing and non-printing areas are on the same plane. Modern offset lithography uses oil-based inks in the form of heavy paste that is metered to the plate cylinder by a train of inking rollers. Another group of rollers applies a thin film of water to the water-receptive areas. Aluminum alloy printing plates are mounted onto a plate cylinder similar to that used in lithography. The ink image trans-fers from the plate to the blanket roll. An impression cylinder provides the

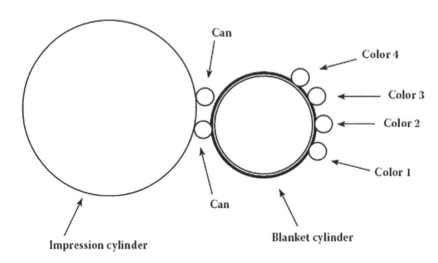

Can

Color 4

Color 3

Color 2

Color 1

Can

Impression cylinder

Blanket cylinder

FIGURE 8.14
Offset letterpress (dry offset).

backing needed to develop slight pressure needed for transferring ink from the blanket to the substrate. UV lithographic printing is the oldest and largest of the energy-curable printing technologies. It is divided into sheet-fed and web printing.[30]

Offset lithography is the most common paper printing process, used for folding carton stock and for can and bottle labels. Most lithographic presses are sheet fed, although there are some web-fed presses. A particular application of offset lithography is the printing of flat metal sheets, which are then formed into metal containers.[29] A schematic of a lithographic printing station is shown in Figure 8.15.

Sheet-fed lithographic printing with UV curable inks is the largest UV ink printing process, followed by UV web lithography, UV screen printing, UV flexography, UV letterpress, and UV inkjet. UV lithographic printing on sheet-fed presses is used mainly for various types of retail folding cartons, commercial printing, metal cans, labels, and food packaging folding cartons.[30]

8.5.5 Rotogravure Printing

Gravure printing separates printing from nonprinting areas by an engraved pattern that is chemically etched or mechanically cut into the surface of a

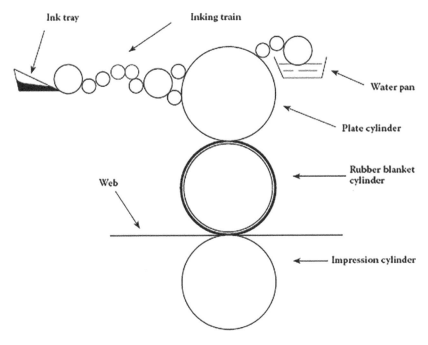

FIGURE 8.15
Lithographic printing station. (Courtesy of Foundation of Flexographic Technical Association.)

hollow metal cylinder. Gravure cylinders are made from copper-plated steel. The printing pattern can be photographically imposed onto the cylinder surface and chemically etched, or it can be imposed by a laser engraved with a stylus.

Normal gravure printing is done from rolls and are in web form. The entire surface of the gravure cylinder is flooded with low-viscosity ink and then wiped clean with a straight-edged doctor blade. The ink remains inside the recessed cell pattern. The substrate to be printed is nipped between the impression roll and the gravure cylinder, and the ink is deposited on the substrate. Gravure printing is used for large-volume applications, such as labels, cartons, carton wraps, and flexible packaging materials.[28] A rotogravure printing station is shown schematically in Figure 8.16, and an example of a pattern on the gravure cylinder is shown in Figure 8.2 (see Section 8.1.1).

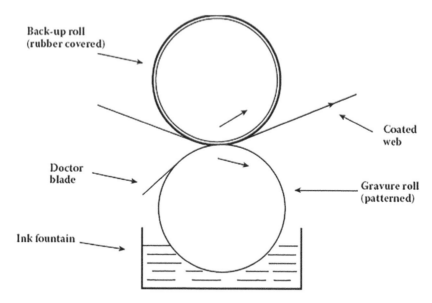

FIGURE 8.16
Rotogravure printing station.

Besides printing and decoration, flexographic, gravure, and lithographic plates as well as silk screens are used in resist chemistry, which is widely employed in the production of printed circuit boards. A resist is a material that will resist solvent attack. A negative resist is a material that, upon exposure to light cross-links, becomes less soluble in a solvent system that would dissolve nonirradiated material. A positive resist is a material that becomes more soluble after irradiation due to depolymerization.[31] A typical printing line with UV curing lines is shown in Figure 8.17.

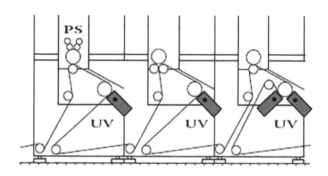

FIGURE 8.17
Printing line with UV curing lamps (PS-printing station, UV-UV curing lamps). (Courtesy of IST Metz.)

8.6 Rapid Prototyping

Rapid prototyping or stereolithography or three-dimensional object curing is a photochemical process used to produce solid three-dimensional objects such as models, masters, or patterns with any shape directly from a design generated on a computer. The computer is used to control the illumination system that builds up the object, usually by a polymerization process. An important part of the design is that the object is sectioned into many small slices so that the polymerization process can be used to build up the object by adding slice upon slice. In the most frequently used process, a continuous wave laser is used to trace out and polymerize a resin in the shape of a slice and to fill in within this area as appropriate. When this process is completed, the next slice is assembled on top of the previously produced slice.[32,33] The objects are thus built up in thin layers of thickness from 0.0005 to 0.020 in. (0.013 to 0.51 mm). The object created by this process is then usually subjected to flood irradiation using a medium mercury lamp to complete the polymerization process. The essential components of this process are[33]:

- A computer with a computer-aided design (CAD) or magnetic resonance imaging (MRI) software package
- A laser to provide light energy
- A mirror system for directing or focusing light energy
- A movable platform located in a reservoir (bath)
- A reservoir containing a photoreactive formulation

The software sends location information to the laser/mirror system, which then directs light energy to specific regions of a thin layer of photoreactive chemical (UV curable formulation) present on a movable platform in the reservoir. Simultaneously, the information package directs the system to lower the platform slowly with the cured slice of the object into the reservoir.[34] The resin systems used in this process are mainly based on methacrylates and acrylates[35] and epoxies.[36] The selection of the resin system depends on the required physical properties.

8.7 UV Powder Coatings

The fundamental benefit of UV powder coatings is that they can be processed at considerably lower temperatures than traditional thermoset powders. While ordinary thermoset powders require temperatures in the range

TABLE 8.7

Comparison of Heating and Curing Methods for Thermosetting and UV-Curable Powders

Thermosetting Powders			
Heating/Curing	Temperature, °C (°F)	Total Curing Time, Minutes	Substrate
Convection	140–220 (285–430)	30–15	Metal
Infrared +convection	140–220 (285–430)	25–10	Metal
Infrared	160–250 (320–480)	15–1	Metal
Induction	240–300 (465–570)	<1	Metal
UV curable powders			
Heating	Curing		
IR/Convection (1–2 minutes)	UV (seconds)	90–120 (195–250)	3–1	Metal, wood, plastics, other

of 350°F–400°F (177°C–204°C), commercial UV curable powders require temperatures that are not higher than 220°F–250°F (104°C–121°C).[37] A comparison of the two types is shown in Table 8.7.

In contrast to the conventional powder coating, the melt and cure processes of UV curable powders are decoupled, so that while heat is used to flow the material, it does not cure the powder, and the cross-linking occurs only after the exposure to UV light. Because the temperature required for melting the UV curable powders is in the range of 175°F–250°F (80°C–121°C),[38] where much less heat is required, and it is needed only long enough to melt the powder but not to cure it. Cross-linking by the UV light is nearly instantaneous; thus, the process is finished in a fraction of the time required for conventional powders.

8.7.1 The Chemistry of UV Curable Powders

To be suitable for radiation curable powder coatings, materials have to contain unsaturated double bonds. They may contain acrylic or methacrylic unsaturations, but nonacrylate systems can also be used.[39] The latter is based on the same chemistry as a commercial nonacrylate 100% reactive liquid system. The binder consists of two components: a resin and a coreactant. The resin is an unsaturated polyester polymer in which maleic acid or fumaric acid is incorporated, and the coreactant is polyurethane containing vinyl ether unsaturation. The UV curing of the binder is based on the 1:1 copolymerization of electron-rich vinyl ether groups with electron-poor maleate or fumarate groups.[40] The system contains a stoichiometric balance of maleate (ME) and vinyl ether (VE) groups. MA/VE binder systems offer the advantage that under controlled conditions, the components do not homopolymerize, and the UV curable powder coating is thermally stable. This means that the time allotted for flow can be increased as needed. This is in contrast to systems containing acrylic and methacrylic unsaturations because they can react under the influence of heat alone.[39] The thermal stability of the MA/VE systems ensures good flow and smooth films.[40] However, since they are applied to substrates with distortion temperatures often under 212°F (100°C), the melt viscosity has to be adjusted accordingly.[41] Another system is based on the photoinduced cationic polymerization of solid bisphenol A-based epoxy resins in the presence of nucleophiles as chain transfer agents.[42]

8.7.2 Material and Substrate Preparation

Since the binder systems are solid at room temperature, they can be produced by the existing methods used for powder coatings.[39] Solid resins, pigments, photoinitiators, and other additives are premixed, then melted and dispersed in an extruder at 100°C–130°C (212°F–266°F). The molten blend is then squeezed into a thin ribbon between chilled rolls. This ribbon is further cooled to near room temperature on a water-cooled cooling belt. The cooled ribbon is broken first into flake and then ground into a fine powder ready for use. The process is illustrated in Figure 8.18.

During the fine grinding process, the final process temperature may reach 30°C–35°C (86°F–95°F). The powder must remain free flowing at these temperatures and not soften or clump together. It is recommended to store the powder coating below 22°C (72°F), but it often happens that the containers with the powder coatings may reach 40°C (104°F) during storage and shipping.

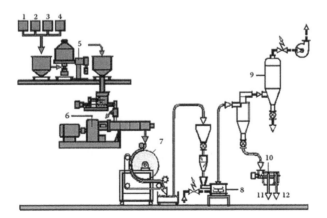

FIGURE 8.18
Process for preparation of powder coatings. (Courtesy of Coperion Compounding Systems.)

To meet these storage recommendations, a powder must not soften below 80°C (176°F) yet must flow and level to a smooth film at 110°C (230°F). Resins, which generally have sufficiently low melt viscosity below this temperature, have a low T_g and consequently low softening points. Powders from such resins could not be stored at 30°C (86°F) and probably not even at 22°C (72°F) without problems. Thus, the melting range for UV coating powders, 110°C–120°C (86°F–248°F), is dictated only by the manufacturing and storage conditions.[43] Relative humidity during storage should not exceed 75%.[44]

Almost any substrate that will be powder coated, whether it is metal, medium-density fiberboard (MDF), or some other material, requires some kind of surface treatment. Surface preparation (pretreatment) of wood and more specifically, MDF preparation, consists of sanding, removal of contaminants, and board conditioning. MDF boards that will be coated by UV curable powders should have a moisture content of 49%. It is also a common practice to preheat MDF boards prior to application of the UV powder.[44]

Metal surfaces are always cleaned by washing and usually treated by a phosphate primer. Conductive plastics are washed and dried and then coated without any further treatment. Nonconductive plastics are made conductive by applying a primer coat or treated by plasma or flame.[45]

8.7.3 Powder Coating Application

The coating is applied electrostatically. The powder is pumped by air through a gun, which generates an electric field and imparts a static charge to the powder particles. These charged particles are attracted uniformly to a grounded part. The part has to be conductive or coated by a conductive primer coat so that it can be grounded (see Section 8.7.2).

The powder is supplied to the gun by a powder pump at a uniform and consistent rate. There are two types of powder pumps: ejector type, using the Venturi principle, and auger pumps, which use positive displacement. A corona charging system is most widely used to electrostatically charge the powder. The electrostatic spray gun has several functions[46]:

- To shape and direct the flow of powder.
- To control the pattern size, shape, and density of powder.
- To impart the electrostatic charge to the powder.
- To control the deposition of powder on the part.

Powder application is shown schematically in Figure 8.19.

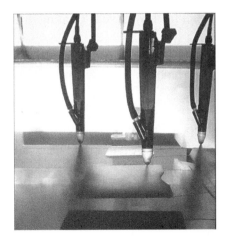

FIGURE 8.19
Application of UV curable powder onto MDF parts. (Courtesy of Nordson Corporation.)

The coated part is then exposed to infrared (IR) heat. The powder coat melts and forms a smooth and uniform film, which usually can take as little as only a few seconds to minutes, depending on the heat surface and the nature of the substrate. The molten film is then held at temperature until maximum flow and leveling are achieved and then immediately exposed to a UV lamp while still molten. Cure usually takes less than 2 seconds.[44]

The heating phase for melting the powder is crucial to the final surface quality. The heat-up should be fast but uniform. The ovens used to melt the powder on the part can be convection, infrared, or combination (hybrid) ovens. Infrared (IR) or combination (IR/convection) ovens are most widely used. IRM (medium-wave infrared adiation) is preferred because it efficiently heats the surface without the rise in the core temperature of the part. Surface temperatures are usually in the range of 100°C–120°C (212°F–250°F). A typical melting process cycle is approximately 3 min, depending on the nature of the substrate.

An example of a line for curing UV powders is shown in Figure 8.20.

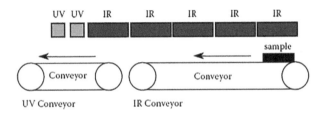

FIGURE 8.20
Principle of curing UV powder coating.

8.7.4 Substrates Suitable for UV Powder Coating

Although UV powder coatings can be applied to a variety of substrates, including metals and plastics with high melting temperatures, their primary advantage is for coating of substrates, which are[46]:

- Temperature-sensitive materials
- Preassembled components
- Dense materials with highly pronounced heat sink characteristics

Plastics: Several types of plastics, such as polystyrene, high-density polyethylene (HDPE), acrylonitrile butadiene styrene (ABS), resin/polybutylene terephthalate (PBT), polyamide, polycarbonate, and polypropylene oxide.

Wood and wood composites: Solid wood as well as particleboard and medium-density fiberboard (MDF) are suitable substrates for powder coating. They tend to emit volatiles such as entrapped air and moisture, and therefore require curing temperatures below 100°C (212°F) and preferably not higher than 80°C (144°F). A typical curing procedure for MDF is shown in Table 8.8.

Preassembled parts: Preassembled parts may contain components made from heat-sensitive substrates (e.g., plastic inserts, wiring). In contrast to the conventional method, where individual components are coated prior to assembly to avoid heat damage, UV powder coating permits the manufacturer to fully assemble a part prior to coating.

Large-mass parts: Conventional coating of large-mass parts requires exceedingly long times in convection ovens. A considerable reduction of times required for cure is to melt the powder with IR and use fast UV-initiated free radical curing reactions.

TABLE 8.8

Typical Curing Conditions for UV Powder Coatings on MDF

Process Variable	Type of Coating		Note
	Clear	Pigmented	
Melt temperature, °C	100–120	100–120	Medium wave IR,
Melt temperature, °F	212–250	212–250	convection or combination ovens
Melt time, seconds	120	120	Typical conditions
Type of UV lamp	Mercury, LED	Gallium doped mercury	Microwave or arc lamps
UV dose, mJ/cm^2	1,000	1,500–4,000	Light bug IL-390

Source: Bayards, R. et al., *RadTech Report*, Vol. 15, No. 5, Sept/October 2001, p. 15 (Modified)

8.7.5 Industrial Applications

The first commercial application was the coating of a fully assembled electric motor.[47] The motor is assembled from parts of different sizes and thickness. Conventional powder coating flows and cures easily on thin and small parts, but the heavier parts may not ever reach the required melting temperature or cure completely even after extended periods of time. This technology is readily applicable to other products, such as wooden furniture with metal hardware.

Currently, the UV powder coating technology is used on a large scale for the following products[48–52]:

- MDF boards for furniture applications
- Preassembled metal objects containing heat-sensitive parts
- PVC flooring

Other applications include[37,53,54]:

- Aluminum wheels
- Automotive accessories (hatchback door closers, hydraulic shock absorbers, headlamp lenses, trim pieces, etc.)
- Electric motors
- Gas cylinders
- Hydraulic door closers
- Heavy-gauge parts
- Carbon-fiber-reinforced composites

8.8 Other Applications for UV Curing

8.8.1 Electronics

Electronics is a major application area for UV curing. Radiation curable resins for resists are the backbone of the imaging process to create electronic circuitry (see Section 8.5) but are also widely used as conformal coatings in the encapsulation of electronic components. In the latter application, UV curable adhesives are used to assemble the components,[55] and UV curable coatings are applied over the assembly to protect it from the environment. Epoxy resins appear to be the most widely used materials,[56–58] with silica-filled epoxies also being important.[59] Epoxy vinyl esters with high silica loading (up to 60 wt.%) are also promising because of their fast cure rate and physical properties, akin to the resins currently in use.[60] Fluorinated silicones, acrylated siloxanes, and cyanates have also been used successfully.[55] In order to ensure cure in the shaded areas, dual-curing systems are often used. A secondary cure system, such as moisture, thermal, or oxidative, is used to augment the radiation energy cure.

Smart cards in the form of SIM cards for mobile (cell) telephones contain chip modules, and UV radiation curable epoxies have been shown to offer advantages in the encapsulation of these vital components.[61]

8.8.2 Optical Components and Optoelectronic Applications

8.8.2.1 Optical Fibers

Optical fibers are widely used in the telecommunication industry for the transmission of digital pulses of voice, video, and data. In order to keep losses in signal strength at a minimum, the fiber (usually a doped silica glass) has to be coated with a material of lower refractive index than its own refractive index. This layer is protected by a "buffer," which acts as a cushion. The buffer is encased in one or more protective layers (see Figure 8.21).[62] Both the primary and protective coatings are very often UV radiation curable.

Since the optical fibers are placed into locations that are difficult to access, the coatings are expected to last over 20 years. Tests indicated that UV-cured protective coatings would be mechanically stable up to 100 years under ambient conditions.[63] Radiation curable adhesives can also be used for end-to-end splicing, termination of bundles, construction of optical sensors, and other areas in the optical field.[64]

8.8.2.2 Other Optical and Optoelectronic Applications

Radiation curing is an important part in the manufacture of compact disks and digital versatile disks (DVDs).[65] In the production of *compact disks*, the replication and formation of the protective coat may be carried out with the use of UV curing.[66] The coatings are commonly based on methacrylates.

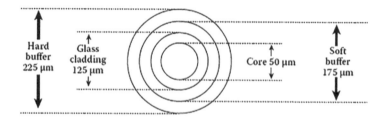

FIGURE 8.21
Schematic of the cross section of an optical fiber. (Adapted from Cutler Jr., O.R., *RadCure '84*, Technical Paper FC84–1022, Atlanta, GA, September 10–13, 1984.)

All *digital versatile disks* are made by cementing two disks together with a clear, UV curable adhesive.[65] The light required has to pass through one of the disks, and consequently, the choice of an initiator having the right absorption properties and of the type of lamp with sufficient power are of great importance.[67]

Other applications are in the manufacture of optical lenses,[68] holograms,[69] and multilayer organic light-emitting devices.[70]

8.9 Automotive Applications

8.9.1 OEM Applications

A considerable amount of work has been done to develop a UV curable system of coatings for OEM applications. The basic requirement is that binders and reactive monomers have excellent weathering resistance. In many specific applications of pure UV curing systems, even in the case of clear coats, surface cracks occurred, depending on curing and storage conditions after curing, due to the volume shrinkage gradient, especially on black and dark base coats. Another shortcoming of pure UV curable systems is the occurrence of shadow areas with a lower state of cure.[71] The solution proposed was a dual-curing system that would include a second cross-linking chemistry—such as OH-functional resins like acrylic acrylates with conventional cross-linkers, for example, isocyanates or acrylates—with NCO functionality or conventional curing systems, such as one-pack alkyd melamine.[71]

In spite of all the efforts mentioned above, no cases of wide application of UV/EB technology are known at this writing. As pointed out earlier, there are many difficulties to do that. The main issues the automotive industry is facing when considering this technology are[72]:

- Standard UV equipment and clear coats cannot ensure curing of coating in the shadowed areas.
- The use of high-intensity, short-wavelength UV lamps, which are known to cure clear coats in seconds, raises concerns about worker safety.
- Many of the clear coats that would be possibly suitable for coating and curing entire vehicle bodies contain free isocyanates and would not be accepted in an assembly line due to health and safety concerns.

On the other hand, there are many potential advantages of using UV or EB technology, such as substantial reduction of energy consumption, CO_2 production, and VOC emissions.

There are great efforts to develop not only clear coats but also primers and base coats that would be curable by UV or EB.[72] The current approaches to UV/EB technology are:

- Clear coats without isocyanates.
- Cationic and photolatent base clear coat formulations curing in shadowed areas.
- The use of a plasma chamber that creates a diffuse light reaching all parts of painted vehicle.[72]
- Cure by electron beam; could apply to primers, base coat, and clear coat.

Even if all the chemistry and technology issues were solved, there would still be questions about the size of the equipment, nitrogen blanketing, and safety measures.

Other automotive applications involve electronic or electrical parts, such as sensor switch encapsulants; conductive inks for the rear-window defroster; motor balancing compounds; windshield wiper motor sealing adhesives; component-marking inks; tacking adhesives; screen-printed membrane switches; headlamps using special UV curing adhesives or conformal coatings; metallization prime coat; lens reflector adhesive (see Sections 8.2 and 8.8.1) and protective hard coats[73]; wheels, motors, and other using UV curable powders (Section 8.7.5); and tail lens assemblies.

One of the lesser known automotive applications is the use of clear coat over molded-in-color body side moldings. The coating has to be very tough and flexible to maintain adhesion without cracking from objects striking the molding. The advantages of UV coatings in this application are the speed of cure (small process footprint) and superior performance properties.[74]

Sheet molding compound (SMC) is a composite material that has been used as an alternative to steel for more than 30 years. SMC consists of a glass fiber-filled unsaturated polyester and vinyl ester resin that has been compacted into a sheet.[75] These sheets are then placed into compression molds and formed into body panels, and these panels are then coated mainly by dual-cure coatings.[76]

Other possible applications are clear coats in automotive interiors (dashboard instrument screen-printed and topcoats for gloss control, interior vinyl clear coat with excellent mar resistance and low gloss, wood grain printing and topcoats for interior laminates), on alloy wheels and wheel covers and under the hood parts,[77] and UV curable paint over physical vapor deposition (PVD) surfaces of various parts.[78] Other examples are:

- Graphics and identifications (deep cure automotive name plates, manufacturers' logos on glass, battery labels, screen-printed oil filter housings, screen-printed windshield washer instructions, etc.)

- Miscellaneous functional applications (black UV coatings for automobile windows, door handle gaskets, UV sealant for airbag explosive cartridges, abrasion-resistant topcoat for metallized plastics, sealers and topcoat for customized van and RV wood components, and more)

8.9.2 Automotive Refinish Applications

UV curing of automotive refinish coatings represents a powerful tool providing a fast curing process and improving the productivity of paint shops. The development of a UV curing refinish primer system has been an important part of the technology. The other important factor has been the equipment based on flashlight technology. This equipment (a xenon lamp) generates UV flashes within a very short time interval with high energy.[71] Also, a homogeneous UV light distribution is realized to cure three-dimensional objects. This procedure is particularly useful for spot repairs with a cure time of less than 60 s. In automotive refinish applications, the typical buildup of the paint layers comprises a putty, primer, base coat, and clear coat, depending on the type and extent of the body damage. In comparison with two-component (2K) urethane and epoxy systems, the UV materials cure considerably faster, and being only one-component coatings, they are ready to use and have practically an unlimited pot life. Moreover, the UV curable primer system can be formulated with much higher solids, i.e., with a considerably lower VOC content,[79] which allows it be compliant with some of the most stringent local VOC regulations. Because of the low VOC content, the risk of encountering pop and gloss dieback is minimal. Additional advantages of UV curable systems are energy savings because UV lamps consume a fraction of the energy required for forced-air baking, and the products are safe to use when following the recommended safety precautions as specified by the product manufacturers.[80,81]

8.9.3 3D Printing

3D printing or additive manufacturing is a process of making three-dimensional solid objects from a digital file. The creation of a 3D printed object is achieved using additive processes. In an additive process, an object is created by laying down successive layers of material until the object is created. Each of these layers can be seen as a thinly sliced horizontal cross section of the eventual object. 3D printing is the opposite of subtractive manufacturing, which is cutting out/hollowing out a piece of metal or plastic with, for instance, a milling machine. 3D printing enables you to produce complex shapes using less material than traditional manufacturing methods. It encompasses many forms of technologies and materials as 3D printing is being used in almost all industries you could think of. It's important to see it as a cluster of diverse industries with a myriad of different applications. There's no question that 3D printing has moved beyond making prototypes and one-offs. Manufacturers and producers are using 3D printers for fabricating parts at various points in the production line, including manufacturing tools, jigs, and fixtures; fit tests; and of course *end-use parts*. The transition to production applications is not just about improving the hardware of 3D printers to the point of replacing legacy manufacturing equipment; manufacturing requires a lot of coordination across large ecosystems, so 3D printing companies are learning to navigate those channels to increase adoption.

Six types *of* materials can be currently used in additive manufacturing: plastics, metals, concrete, ceramics, paper, and other. Materials are often produced in wire feedstock, so-called 3D printer filament, powder form, or liquid resin. All seven previously described 3D printing techniques cover the use of these materials, although polymers are most commonly used and some additive techniques lend themselves toward the use of certain materials over others.

A few examples:

- Dental products
- Eyewear
- Architectural scale models and maquettes
- Prosthetics
- Movie props

Companies have used 3D printers in their design process to create prototypes since the late 1970s. Using 3D printers for these purposes is called *rapid prototyping*. Besides rapid prototyping, 3D printing is also used for *rapid manufacturing*. Rapid manufacturing is a new method of manufacturing where businesses use 3D printers for short-run/small-batch custom manufacturing; UV-LED printers use a light-emitting diode (LED) in the print head instead of a laser, as in the case of laser printers, thus the name LED. However, in a UV-LED printer, the lamp emits energy onto specially formulated ink containing photoinitiators, which cures the ink directly onto the substrate.

The light from the flat LED panels (arrays) shines directly, in a parallel fashion, onto the build area. Because this light is not expanded, pixel distortion is less of an issue with LED printing. This means that the print quality of an LED printer depends on its light density. The more pixels it has, the better the print quality. Resin 3D printing is also known as vat polymerization. Vat polymerization technologies involve a photosensitive resin cured by a light source to produce solid layers and, eventually, whole parts. The resin is contained within a vat, or tank, and is cured against a build platform, which slowly rises out of the tank as the part is formed. Ultimately, not all 3D printing processes allow for UV curing, but there are a few, namely stereo lithography (SLA), digital light processing (DLP), and inkjet printing being the earliest forms of 3D printing to use in the process.[82] The subject of formulating for 3D printing, components, and issues are the focus of an article by Viereckl.[83]

8.10 Production of Composites by UV Radiation

8.10.1 Dental Applications

UV radiation curable dental compositions have been used since the early 1970s.[84,85] They consist of photocurable resins with silica filler and offer many advantages: they are ready for use on demand, offer an extended working time, and have an absence of air bubbles, a high polymerization rate, and good color stability. The silica filler provides mechanical strength and reduces shrinkage. However, the formulations have to meet several stringent requirements: The use of light with a wavelength shorter than 330 nm is unacceptable, since it may cause tissue damage and requires that the patient and the operator wear special protective wear. Therefore, photoinitiators for longer wavelengths are used (e.g., camphorquinone).[83] The material should be polishable, opaque to X-rays, and stable toward saliva, beverages, etc. Methacrylates are used as the base resin, in spite of their slow cure rate when compared to acrylates, which are less favorable because of their toxicological properties.

Fissure sealants are clear, low-viscosity, photocurable liquids, based on a mixture of tetrafunctional and difunctional methacrylates. Other dental photocurable materials are cements, adhesives, and denture base resins (Andrzejewska, E., private communication, June 10, 2002).[86,87]

Glas-ionomer cements are acid-based materials (using, e.g., polyacrylic acid), whose setting reaction involves neutralization of the acid groups by powdered solid bases (calcium fluoroaluminosilicate glasses). Resin-modified glass ionomer cements are hybrid materials prepared by the incorporation of polymerizable components such as 2-hydroxyethyl methacrylate.[88]

8.10.2 Other Composite Applications

UV curing is used in several specialized applications in composite materials. Most of the radiation curable composite materials are processed by electron beam or microwave energy, mainly due to the difficulties of getting the irradiating light to penetrate thick sections containing particles or fibers, which are often opaque to light.[89] The examples of UV-cured composite materials include cure of glass and other fibers impregnated by photocurable resins in the production of coated filaments and in the production of prepreg sheets for waterproofing buildings and in filament winding.[90] Resin containing carbon fibers were reported to be cured with UV light.[91]

8.11 Hydrogels

Hydrogels are three-dimensional, cross-linked hydrophilic polymeric structures able to absorb large quantities of water. They are synthesized by polymerization of hydrophilic monomers. The extent of the reversible swelling and deswelling property of these materials is known to depend on the nature of both intermolecular and intramolecular cross-linking, as well as the degree of hydrogen bonding in the polymer network. Hydrogels are mainly produced by the use of radiation processes, both UV and EB (more in Section 9.5.6). An example of a mature technology of this kind is the production of hydrogel wound dressings, highly absorbent diapers, contact lenses, and similar articles now being produced on a large scale; emerging applications include a hydrogel-based system for anticancer therapy due to local drug delivery and a variety of other medical applications.

Polyurethane hydrogels derived from UV curable urethane prepolymer and hydrophilic monomers were prepared and their properties were evaluated. The urethane prepolymer used in this study contained well-defined hard segments centered with a polyether-based soft segment and end-capped with methacrylate groups. The hydrophilic monomers studied were 2-hydroxyethyl methacrylate (HEMA), *N*-vinyl pyrrolidone, and glycerol methacrylate. Methacryloxypropyl tri(trimethysiloxy) silane (TRIS) was also used in some cases to modify properties. All compositions were UV-cured and formed hydrogels after hydration. The oxygen permeabilities of the hydrogels decreased as the water contents increased, and increased as the TRIS content was increased. The tear strengths and moduli decreased as the water contents of the hydrogels increased. Most compositions studied have higher oxygen permeability and tear strength than poly(HEMA) because of the presence of urethane prepolymer in the composition. The hydrogels are used for a variety of biomedical applications.[92]

Poly(*N*-vinyl-2-pyrrolidone) (PVP) hydrogels produced by high-energy radiation rely on water radiolysis as a primary process leading to cross-links. Conversely, ultraviolet direct irradiation into PVP leads to cross-linking through pyrrolidinone moiety photolysis. However, this process showed to be rather inefficient. However, cross-linking of PVP based on hydrogen peroxide photolysis, therefore mimicking water radiolysis, using UVC (e.g., low-pressure Hg lamp) or UVA radiation sources, is more efficient. The process efficiency and the properties of the hydrogel formed are discussed and compared with those of other methods of hydrogel production. Hydrogels based on PVP prepared by this method find use in biomedical applications.[93]

References

1. Mehnert, R., Pincus, A., Janorsky, I., Stowe, R., and Berejka, A., *UV&EB Curing Technology & Equipment*, Vol. I, John Wiley & Sons Ltd., Chichester/ SITA Technology Ltd., London, p. 159 (1998).
2. Mehnert, R., Pincus, A., Janorsky, I., Stowe, R., and Berejka, A., *UV&EB Curing Technology & Equipment*, Vol. I, John Wiley & Sons Ltd., Chichester/ SITA Technology Ltd., London, p. 169 (1998).
3. Mehnert, R., Pincus, A., Janorsky, I., Stowe, R., and Berejka, A., *UV&EB Curing Technology & Equipment*, Vol. I, John Wiley & Sons Ltd., Chichester/ SITA Technology Ltd., London, p. 171 (1998).
4. Stranges, A., *Proceedings RadTech '94 North America*, p. 415 (1994).
5. Jones, D. T., *Proceedings RadTech '94 North America*, p. 417 (1994).
6. Mehnert, R., Pincus, A., Janorsky, I., Stowe, R., and Berejka, A., *UV&EB Curing Technology & Equipment*, Vol. I, John Wiley & Sons Ltd., Chichester/ SITA Technology Ltd., London, p. 173 (1998).
7. Cohen, E. D., and Gutoff, E. B. (Eds.), *Modern Coating and Drying Technology*, VCH, Weinheim (1992).
8. Zimmerman, T., *Proceedings RadTech Europe '93*, p. 698 (1993).
9. Mehnert, R., Pincus, A., Janorsky, I., Stowe, R., and Berejka, A., *UV&EB Curing Technology & Equipment*, Vol. I, John Wiley & Sons Ltd., Chichester/ SITA Technology Ltd., London, p. 184 (1998).
10. Mehnert, R., Pincus, A., Janorsky, I., Stowe, R., and Berejka, A., *UV&EB Curing Technology & Equipment*, Vol. I, John Wiley & Sons Ltd., Chichester/ SITA Technology Ltd., London, p. 187 (1998).
11. Mehnert, R., Pincus, A., Janorsky, I., Stowe, R., and Berejka, A., *UV&EB Curing Technology & Equipment*, Vol. I, John Wiley & Sons Ltd., Chichester/ SITA Technology Ltd., London, p. 189 (1998).
12. Roesch, K. F., *Proceedings RadTech Europe '93*, p. 940 (1993).
13. Garratt, P. G., *Proceedings RadTech Europe '87*, p. 10 (1987).
14. Roth, A., and Honig, M., *Proceedings RadTech North America '98*, p. 112 (1998).
15. *UV/EB Primer: Inks, Coatings and Adhesives*, RadTech International North America, Bethesda, MD, p. 13 (1995).
16. Des Roches, S., *Adhesives & Sealants Industry*, 12, No. 5, August, pp. 10–14 (2005).
17. Johnston, J., *Pressure-Sensitive Tapes: A Guide to Their Function, Design, Manufacture and Use*, Pressure Sensitive Tape Council, Northbrook, IL, (2003).
18. Benedek, I., *Pressure-Sensitive Adhesives and Applications*, 2nd ed., CRC Press, Boca Raton, FL (2004).
19. Des Roches, S., and Murphy, W., *Adhesives & Sealants Industry*, 13, No. 8, August, p. 26 (2006).
20. Huber, H., Anzures, E., and Aceveda, M., *RadTech Report*, 11, No. 2, March/ April, p. 17 (1997).
21. Nunez, C., McMinn, G., and Vitas, J., *RadTech North America '94 Conference Proceedings*, p. 42 (1994).

22. Mehnert, R., Pincus, A., Janorsky, I., Stowe, R., and Berejka, A., *UV&EB Curing Technology & Equipment*, Vol. I, John Wiley & Sons Ltd., Chichester/SITA Technology Ltd., London, p. 185 (1998).

23. Burga, R., *Proceedings RadTech Asia '97*, p. 192 (1997).

24. Robins, G., in *Radiation Curing of Polymers* (Randell, D. R., Ed.), The Royal Society of Chemistry, London, p. 78 (1987).

25. Caza, M., *Proceedings of RadTech Europe '97*, p. 392 (1997).

26. *UV/EB Curing Primer: Inks, Coatings and Adhesives*, RadTech International North America, Bethesda, MD, p. 17 (1995).

27. Nigam, B., *FLEXO*, No. 9, September, p. 8 (2001).

28. Rasmussen, M. P., *RadTech Report*, 13, No. 2, p. 8 (1998).

29. Soroka, W., *Fundamentals of Packaging Technology*, 2nd ed., Institute of Packaging Professionals, Herndon, VA (1999).

30. Duncan, D., in *RadTech Printer's Guide*, www.radtech.org, p. 65.

31. Roffey, C. G., *Photopolymerization of Surface Coatings*, John Wiley & Sons Ltd., Chichester (1998).

32. *Stereolithography (SLA)*, Tech, Inc., www.techok.com/sla (2000).

33. Narey, B., *Conference Proceedings, Innovation in Plastics: Plastics Technologies for a Global Competitive Edge*, Rochester, September 18–19, Paper J.012 (1990).

34. Koleske, J. V., *Radiation Curing of Coatings*, ASTM International, West Conshohocken, PA, p. 216 (2002).

35. Popat, A. H., and Lawson, J. R., U.S. Patent 6,025,114 to Zeneca Ltd. (2000).

36. Yamamura, T., et al., U.S. Patent 5,981,616 to DSM, NV, JSR Corporation, Japan Fine Coatings Co. Ltd. (1999).

37. Mills, P., *RadTech Report*, 11, No. 6, November/December, p. 16 (1997).

38. Mills, P., Blatter, W., Binder, J., and Mitchell, S., *UV Powder Application Guide*, RadTech International North America, Chevy Chase, MD, Chapter 1, p. 6 (2002).

39. Witte, F. M., de Jong, E. S., and Misev, T. A., *RadTech Report*, 10, No. 5, September/October, p. 14 (1996).

40. Zahora, E. P., et al., *Modern Paint and Coatings*, October, p. 120 (1994).

41. Witte, F. M., et al., *RadTech Europe 1995, Conference Proceedings*, Maastricht, p. 437 (1995).

42. Finter, J., Frischinger, I., Haug, Th., and Marton, R., *RadTech Europe '97 Conference Proceedings*, Lyon, France, p. 489 (1997).

43. Gottschling, P., and Stachyra, Z., *RadTech Report*, 13, No. 4, July/August, p. 42 (1999).

44. Mills, P., *UV Powder Application Guide*, RadTech International North America, Chevy Chase, MD, Chapter 3, p. 15 (2002).

45. Mihalic, S. D., and Mills, P., *RadTech Report*, 13, No. 4, July/August, p. 26 (1999).

46. Drummond, K. C., and Guskov, S. V., *RadTech Report*, 13, No. 4, July/August, p. 22 (1999).

47. Mills, P., *RadTech Report*, 13, No. 4, July/August, p. 31 (1999).

48. Buysens, K., and Hammerton, D. A., *RadTech Report*, 13, No. 4, July/August, p. 18 (1999).

49. Buysens, K., and Jacques, K., *Eur. Coatings J.*, 9, p. 22 (2001).

50. Bayards, R., et al., *RadTech Report*, 15, No. 5, September/October, p. 15 (2001).

51. Heathcote, J., *RadTech Report*, 15, No. 5, September/October, p. 10 (2001).

52. Karlsson, L., and Leach, C., *Industrial Paint & Powder*, 78, No. 4, April, p. 12 (2002).

53. Mills, P., *RadTech Report*, 15, No. 5, September/October, p. 32 (2001).

54. McFadden, B., *RadTech Report*, 11, No. 1, January/February, p. 15 (1997).
55. Davidson, R. S., *Radiation Curing, RAPRA Review Reports*, 12, No. 4, p. 24 (2001).
56. Bittmann, E., and Ehrenstein, G. W., *Plaste u. Kautschuk*, 41, No. 5, September, p. 216 (1994).
57. Bittmann, E., and Ehrenstein, G. W., *ANTEC '96, Conference Proceedings*, Vol. II, Indianapolis, IN, May 5–10, p. 1465 (1996).
58. Sheehan, J., *Pitture e Vernici*, 71, No. 18, November, p. 26 (1995). In Italian and English.
59. Stampfer, S., and Ehrenstein, G. W., *ANTEC '98, Conference Proceedings*, Vol. II, April 26–30, p. 1173 (1998).
60. Baikerikar, K. K., et al., *Conference Proceedings, RadTech North America '98*, p. 712 (1998).
61. Most, R., *European Design Engineer*, April, p. 58 (2000).
62. Davidson, R. S., *Radiation Curing, RAPRA Review Reports*, 12, No. 4, p. 18 (2001).
63. Cutler Jr., O. R., *RadCure '84*, Technical Paper FC84–1022, Atlanta, GA, September 10–13 (1984).
64. Murray, R. T., and Jones, M. E., *Command Cure Precision Cements in Optics*, Technical Information Brochure, ICI Resins-US (1993).
65. Ledwith, A., in *Photochemistry and Polymeric Systems* (Kelly, J. M., et al., Eds.), Royal Society of Chemistry, Cambridge, UK, p. 1 (1993).
66. Kitans, R., and Amo, M., EP 962301 to Kitano Engineering Co. Ltd. (1990).
67. Skinner, D., *Proceedings RadTech North America '98*, Chicago, IL, p. 140 (1998).
68. Luetke, S., *RadTech Report*, 10, No. 2, March/April, p. 29 (1996).
69. Sponsler, M. B., *The Spectrum*, 13, p. 7 (2000).
70. Braig, T., et al., *Conference Proceedings, ACS Polymeric Materials Science and Engineering*, Vol. 80, Spring, Anaheim, CA, p. 122 (1999).
71. Maag, K., Lenhard, W., and Loffler, H., *Prog. Org. Coat.*, 40, p. 93 (2000).
72. Seubert, C. M., and Nichols, M. E., *RadTech Report*, 22, No. 6, November/December, p. 27 (2008).
73. Slocum, S., and Mordhorst, S., *Radtech Report*, 22, No. 6, November/December, p. 40 (2008).
74. Joesel, K., *Product Finishing—UV Coating Technology Supplement*, p. 16s (2004).
75. Mallick, P. K., in *Composite Materials Technology* (Mallick, P. K., and Newman, S., Eds.), Hanser Publishers, Munich, 1990, p. 27.
76. Joesel, K., *Product Finishing—UV Coating Technology Supplement*, p. 18s (2004).
77. Joesel, K., *Product Finishing—UV Coating Technology Supplement*, p. 21s (2004).
78. De Stephani, J., *Product Finishings*, August (2009).
79. Stropp, J. P., et al., *Progr. Org. Coat.*, 55, p. 29 (2006).
80. *UV Cure Automotive Refinishing Products*, RadTech International North America, www.radtech.org.
81. *Body Shop Business*, May (2005).
82. Rolland, J., "How can a new UV-cured chemistries and materials enable expanded applications in 3D printing?" *UV + EB Technology*, 4, No. 4, p. 12 (2018).
83. Viereckl, J. A., et al., "Formulating for 3D printing: Constraints and components for stereolithigraphy," *UV + EB Technology*, 4, No. 4, p. 52 (2018).
84. Linden, L.-A., in *Radiation Curing in Polymer Science and Technology*, Vol. IV (Fouassier, J. P., and Rabek, J. F., Eds.), Elsevier Applied Science, London, p. 387 (1993).

85. Anon., *Revista de Plasticos Modernos*, 78, No. 519, September, p. 287 (1999). In Spanish.
86. Woods, J. G., in *Radiation Science and Technology* (Pappas, S. P., Ed.), Plenum Press, New York, Chapter 9, p. 333 (1992).
87. Linden, L. A., in *Polymeric Materials Encyclopedia* (Salamone, J. C., Ed.), CRC Press, Boca Raton, FL, p. 1839 (1996).
88. Andrzejewska, E., Andrzejewski, M., Socha, E., and Zych-Tomkowiak, D., *Dent. Mater.*, 19, p. 501 (2003).
89. Davidson, R. S., *Radiation Curing, RAPRA Review Reports*, 12, No. 4, Rapra Technology Ltd., p. 19 (2001).
90. Graff, G., *Modern Plastics International*, 26, No. 9, September, p. 38 (1996).
91. Muranaka, K., U.S. Patent 5, 591,784 to Three Bond Co. Ltd. (1997).
92. Lai, Y.-C., and Baccei, L. J., *J. Appl. Polym. Sci.*, 42, p. 3173 (1991).
93. Fechine, G. J. M., Barros, J. A. G., and Catalani, L. H., *Polymer*, 46, p. 283 (2005).

9

Applications of Electron Beam Radiation

The chemical reactions of monomers, oligomers, and significant changes in properties of polymers induced by ionizing radiation discussed in previous chapters can be used for a variety of practical applications. Besides the already mentioned advantages in clean and safe technology, the almost instant conversion and excellent control of dosage and penetration depth electron beam processing found its way to a variety of industrial applications, such as wire and cable insulations; tire manufacturing; production of polymeric foams, heat-shrinkable films, and tubings; curing of coatings, adhesives, and composites; as well as printing. Large-scale industrial applications of ionizing radiation started in the late 1950s, when Raychem introduced the production of polyethylene heat-shrinkable tubing and W. R. Grace started to manufacture polyolefin packaging. At about the same time, Goodyear and Firestone initiated investigation of modification of rubber compounds by electron beam irradiation for tire applications.[1]

The equipment used must be matched to the given process. For example, for wire and cable, as well as for precuring of tire components and processing of rubber products, and polyolefin foams, where greater electron penetration is required, accelerators with energies ranging from 0.5 to 5 MeV and power rating less than 200 kW are used.[2] Coatings, adhesives, printing, and thin films can be processed by low-energy accelerators with energies typically in the 100–500 keV range or even less. Some of the major industrial applications using electron beam processing of polymeric systems are shown in Table 9.1.

9.1 Electron Beam Process in Wire and Cable Technology

More than 30% of the industrial electron beam processors in the world are used in the radiation cross-linking of wire and cable insulation,[3] and this application is growing.

The conductors produced by the wire and cable industry can be roughly divided into two basic groups: wires and cables. The difference between them is that a *wire* is a single conductor, and a *cable* is a group of two or more insulated conductors. If there were not any insulation on the two conductors, then it would not be a cable; it would still be a single conductor that would

TABLE 9.1

Major Commercial Applications of Electron Beam Processing of Polymeric Materials

Material/Substrate	Application	EB Process
Polyolefins and polyvinyl chloride, some elastomers	Wire and cable insulation	Cross-linking, 0.4–3 MeV, at approx. 10 kGy or higher
Elastomers	Tire manufacture, improved green strength of components and tire performance	Cross-linking with high-energy electrons
Polyolefins and PVC	Improving thermal stability, uniformity and fine structure for packaging and insulation, rubber sheeting	Cross-linking with high-energy electrons
Wood impregnated with acrylic and methacrylic monomers	No-wear high performance floors for high traffic areas	Polymerization by electron beam
Polyolefins	Heat-shrink films and tubing	Cross-linking
Polymeric films, metallic foils, paper, metal, wood	Curing of adhesives, coatings and inks, printing	Low energy processing in the 100–500 keV range at 0.1–0.2 MGy
Polytetrafluoroethylene	Degradation into low molecular weight products ("micropowders") used as additives to coatings, lubricants, and inks	High-energy irradiation at 200–400 kGy

classify as a wire. There are four basic categories of wire and cable products. These categories consist of single-conductor, multiconductor, twisted pairs, and coaxial cable. The most widely used conductor materials used are copper and aluminum.

The two main manufacturing steps in wire and cable production are extrusion and vulcanization (cross-linking). *Extrusion* in this case is done by a process referred to as cross-head extrusion. The basic procedure includes pulling of the wire or cable to be coated at a uniform rate by a *crosshead die* where it is covered with the molten plastic or hot rubber compound. The crosshead extrusion process is carried out by using equipment in the manufacturing line. In this process, *primary insulation* is defined as the polymeric material applied directly onto the metal wire/cable to isolate the metal electrically. *Jacketing* (or *sheathing*) refers to covering on a wire or a group of wires with an insulating coating or jacket for non-electrical protection. Jackets are usually put onto primary wires. The cooling of the extruded insulation or jacket is done in air or water. Various polymers are used in wire coating applications by the crosshead extrusion process. The characteristics of these materials, which make them ideal for this purpose, are their flexibility; desirable electrical properties; ability to resist to chemical, mechanical, and environmental damage; and durability.

The most widely used process, particularly in North America, to vulcanize (cross-link) rubber W&C compounds is the *continuous vulcanization (CV operation)*. The production unit consists essentially of an extruder attached to a jacketed curing tube in which high-pressure steam is confined. The wire or cable with the compound applied by extrusion discharges directly from the extruder head into the steam tube through which it is conveyed under tension. Steam pressures of 1.4 MPa (200 psi) and higher are generally used, and tube lengths may be 60 m (200 ft) or longer. The heated tube is often divided into several separately controlled heat zones, which can be adjusted up to temperature as high as 450°C (842°F). Compounds for continuous vulcanization are designed to cure in seconds at elevated temperatures, and their handling and processing are somewhat more critical than in conventional rubber manufacturing

The largest volume of polymeric materials used for wire and cable insulation are thermoplastics, namely, polyethylene (PE) and polyvinylchloride (PVC)[4] and to a lesser degree elastomeric compounds. The main reason for the prevalence of PE and PVC in wire and cable insulation is their easy processing and relatively low cost. However, their main disadvantage is that their physical properties, such as plastic flow at elevated temperatures, environmental stress cracking, poor solvent resistance, and low softening temperatures[3] cannot always meet demands imposed on them by modern applications. Cross-linking of these materials improves their toughness, flexibility, and impact resistance, resistance to solvents and chemicals, as well as their service temperature.[5,6]

There are essentially three cross-linking processes used in the wire and cable industry. Two kinds of chemical cross-linking, employing organic peroxides or silanes, respectively, are still frequently used for the improvement of PE and other polymer-based insulations for wire and cable.[7,8] However, the advantages of radiation cross-linking over the chemical process have brought about its steady growth of about 10%–15% annually[9] since the mid-1970s, when high-energy and high-current electron accelerators became widely available for radiation processing.[10,11]

The most common compound for peroxide cross-linking of PE in the wire and cable industry is dicumyl peroxide (DCP), which decomposes at 120°C–125°C (248°F–257°F), generating free radicals needed for the process plus by-products (carbinol, acetophenone, and methane).[7] The production line for continuous vulcanization of wire and cables by peroxide must be about 200 m (61 ft) long.[12] The plant space required for radiation cross-linking using electron accelerators is considerably smaller. Moreover, energy consumption for the radiation process is much lower than that for the chemical cross-linking process.[13] As pointed out earlier, the dose required for the desirable degree of cross-linking is determined by the electron current of the accelerator and by the wire or cable speed; therefore, the process control is quite simple.[14] Depending on the source (most commonly steam) and the size of the product, regulation of the temperature of the vulcanization line,

TABLE 9.2

Comparison of Radiation and Peroxide Cross-Linking of Wire and Cable Insulation Materials

Property	Radiation Cross-Linking	Peroxide Cross-Linking
Energy consumption	Low	High
Line speed, m/min (ft/min)	Fast, up to 500 (1640)	Slow, max. 200 (655)
Processing	Offline	Online
Factory space	Relatively small	Relatively large
Product sizes	Wide variety	Fixed
Process control	Current and line speed	Heat flow
Insulation materials[a]	LDPE, FRLDPE, HDPE, FRHDPE, PVC	LDPE
Maintenance cost	Low	High
Start-up	No scrap	100 m (330 ft) scrap at start
Voltage rating, kV	Up to 5	50
Capital cost	High	Moderate

[a] LDPE = low-density polyethylene, FRLDPE = flame-retardant LDPE, HDPE = high-density polyethylene, FRHDPE = flame-retardant HDPE, PVC = polyvinyl chloride.

typically in the range of 200°C–400°C (392°F–752°F), is more complicated than the control of the electron beam current. While chemical cross-linking is applicable only to polyethylene and elastomers, radiation can be used for the cross-linking of PE, PE with flame retardants, PVC, some fluoropolymers, and some elastomers. A comparison of peroxide and radiation cross-linking in the wire and cable applications is shown in Table 9.2.

The silane cross-linking of polyethylene has been used mainly in low-voltage applications. It is a two-stage process, and the curing time is dictated by the time required for the moisture to diffuse thoroughly into the insulation. Consequently, the process is rather slow. Another drawback of this system is the limited shelf life of the silane cross-linking agent and the significant cost of the compound.[15]

Polymeric materials used for electrical insulation have to exhibit not only good electrical properties but also good mechanical properties and good appearance. In addition to reliable electrical performance, thermal stability is important, mainly for miniaturized electronic and electrical equipment. Long-term stability at temperatures in the range of 90°C–125°C (194°F–257°F) is required. The short-term stability determines the resistance of the insulation to melted solder and its cut-through resistance.

The most important performance tests on insulating materials are those for dielectric strength and insulation resistance. Wire and cable products are required to perform over different voltage ranges and at different conditions, such as humid environment and after aging.[16] One of the tests is dielectric breakdown. Dielectric breakdown occurs when electrons detached from the molecule acquire sufficient energy in the electric field to yield secondary ionization and avalanche.[17] For most polymers, the dielectric strength can be as high as 1,000 MV/m. However, under practical conditions, breakdown of polymer insulation occurs at much lower electric field strengths. For example, if the power dissipated in the insulating material raises its temperature enough to cause a thermal stress, the dielectric breakdown occurs at a much lower field strength. Another factor is surface contamination, which can cause many polymeric insulators to break down by tracking on their surface.[18]

Two other important electrical properties must be taken into consideration when polymers are used as insulation for a high-voltage power cable[19] or electronic wires.[20] These are the dielectric constant and the dielectric loss factor, which characterize the energy dissipation in the insulation, the capacitance, the impedance, and the attenuation.

In contrast to metals and semiconductors, the valence electrons in polymers are localized in covalent bonds.[21] The small current that flows through polymers upon the application of an electric field arises mainly from structural defects and impurities. Additives such as fillers, antioxidants, plasticizers, and processing aids of flame retardants cause an increase of charge carriers, which results in a decrease of their volume resistivity.[22] In radiation cross-linking, electrons may produce radiation defects in the material; the higher the absorbed dose, the greater the number of defects. As a result, the resistivity of a radiation cross-linked polymer may decrease.[23] Volume resistivities and dielectric constants of some polymers used as insulations are shown in Table 9.3. It can be seen that the values of dielectric constants of cross-linked polymers are slightly lower than those of polymers not cross-linked.

Abrasion resistance and solder iron resistances are two other important properties required from a wire and cable insulator. These properties are critical for polymers used as sheathing or jacketing for hookup wire. Cross-linking improves both of them.[24]

TABLE 9.3

Volume Resistivity and Dielectric Constant Values of Polymeric Insulating Materials

		Dielectric Constant	
Insulating Material	Volume Resistivity, ohm·cm	Uncured	Cross-Linked
Low-density polyethylene	10^{18}	2.3	2.0
High-density polyethylene	10^{16}	NA	NA
Rigid PVC	10^{16}	NA	NA
Plasticized PVC	$10^{11}-10^{14}$	3.8	3.2
Polypropylene	10^{16}	2.3	NA
EPDM	10^{16}	3.1	3.6

Cross-linkable polymers used for wire and cable insulations are polyolefins, certain fluoropolymers, and elastomers. Among these, radiation cross-linked polyethylene is the most widely used. The radiation cross-linking process of PE has also been the most widely studied.[25–28]

High-voltage PE insulation thicker than 4 mm (0.16 in.) tends to form voids due to heat accumulation, the evolution of hydrogen gas, and discharge breakdown due to accumulation of excess charges[29] when irradiated. To eliminate this problem, the addition of prorads (such as multifunctional acrylate and methacrylate monomers) to reduce required dose[30] is necessary. Some monomers, such as triallyl cyanurate and dipropargyl succinate, are effective in reducing the dose required for cross-linking, but they have an adverse effect on the dissipation factor of the insulation.[31]

PVC, another widely used polymer for wire and cable insulation, cross-links under irradiation in an inert atmosphere. When irradiated in air, scission predominates.[32,33] To make cross-linking dominant, multifunctional monomers, such as trifunctional acrylates and methacrylates must be added.[34,35]

Fluoropolymers, such as copolymers of ethylene and tetrafluoroethylene (ETFE), or polyvinylidene fluoride (PVDF) and polyvinyl fluoride (PVF), are widely used in wire and cable insulations. They are relatively easy to process and have excellent chemical and thermal resistance, but tend to creep, crack, and possess low mechanical stress at temperatures near their melting points. Radiation has been found to improve their mechanical properties and crack resistance.[36]

Ethylene propylene rubber (EPR) has also been used for wire and cable insulation. When blended with thermoplastic polyefins, such as low-density polyethylene (LDPE), its processibility improves significantly. The typical addition of LDPE is 10%. Ethylene propylene copolymers and terpolymers with high PE content can be cross-linked by irradiation.[34]

9.1.1 Equipment and Process

In general, radiation cross-linking of wire and cable insulation requires medium-energy electrons in the range of 0.5–2.5 MeV. The electrostatic electron beam systems using high DC voltage to accelerate electrons have been found to be most suitable.[37] Cross-linking of wire and cable insulation results from the dissipation of the energy carried by fast electrons penetrating into the insulation. As pointed out in Section 2.2, the depth of penetration is governed by the accelerating voltage. For that reason, the equipment used has to have sufficient beam energy, since it defines the maximum thickness of the insulation, which can be penetrated and effectively cross-linked by the electrons.[38,39] The effective depth (optimum penetration) of the electron beam is defined as the thickness in which the exit plane dose equals the incident surface dose. The ratio of the peak dose to the entrance dose is defined as dose uniformity.[37] Because of the nature of the process,[40] the energy imparted to the insulating material is not uniformly distributed. However, the physical properties of the insulation modified by radiation cross-linking are not very sensitive to the variation of the dose in depth. Thus, the heterogeneity of the dose should not be of great concern, and the maximum-to-minimum-dose ratio of 1.5 or less is usually satisfactory.[37]

The volume resistivity of polyethylene is very high, and the electrons that come to rest in the insulation layer cannot be easily removed. Therefore, it is necessary for the electrons irradiating the wire or cable to penetrate through the insulation to ensure that no charge is accumulated in the wire or cable, causing dielectric breakdown.[41] PVC has a lower resistivity than PE; therefore, charges during the irradiation can leak from the insulation through the wire or cable conductor to ground. The penetration depth (or range of electrons) for a given electron beam depends on the thickness of the wire or cable insulation, and the diameter of the conductor can be calculated.[42,43] Usually, the penetration of twice the radial thickness of the insulation is sufficient.[43]

Irradiation cross-linking of small wires and cables with insulation thickness around 1 mm (0.040 in.) is done under a scanner. The wire or cable is strung between two drums to form a figure eight (see Figure 9.1) and is irradiated on both sides. The dose is controlled automatically by a servolink.

Cables with a large conductor cross section (e.g., 150 mm²) and thicker insulation (2 mm or 0.080 in.) require a multiside irradiation.[43–45] An example of a four-sided irradiation using a single accelerator is shown in Figure 9.2,[45]

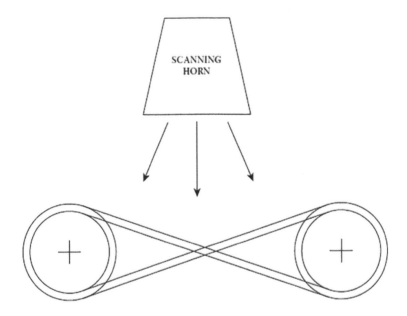

FIGURE 9.1
Schematic of two-sided irradiation of a wire by electron beam. (From *Radiation Processing of Polymers*, Singh, A. and Silverman, J., Eds., Carl Hanser Publishers, Munich, p. 82. With permission.)

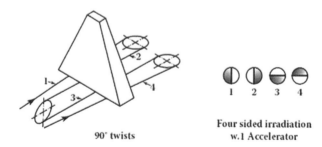

90° twists

Four sided irradiation
w.1 Accelerator

FIGURE 9.2
Schematic of four-sided irradiation of a wire. (From *Radiation Processing of Polymers*, Singh, A. and Silverman, J., Eds., Carl Hanser Publishers, Munich, p. 84. With permission.)

FIGURE 9.3
Actual two-sided irradiation of a wire or cable, the "Figure Eight." (Courtesy of M. R. Cleland.)

where the scanning horn is placed over four wire/cable strands twisted between rollers. The actual two-sided irradiation of a wire or cable is shown in Figure 9.3 and the actual multi-pass wire irradiation fixture is shown in Figure 9.4. The cable speed, the rotation speed of the twisting cable, and the scanner frequency are coordinated by a servomechanism.[45] Some high-frequency accelerators can provide deflected electron beams capable of irradiating the cable from three directions. In such an arrangement, the required

FIGURE 9.4
Actual multi-pass wire irradiation fixture for wire or small tubing. (Courtesy of IBA Industrial.)

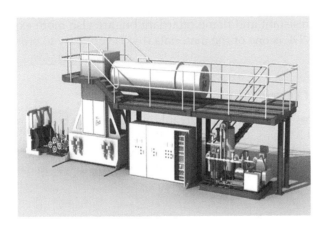

FIGURE 9.5
Modern production EB line for wire and cable. (Courtesy of IBA Industrial.)

electron penetration equals the required thickness of insulation; consequently, electron accelerators of lower energy can be used.[45] The modern EB
production line for W&C is shown in the Figure 9.5.

The doses required for electron beam cross-linking of PE and PVC (compounded with prorads) are 100–300 kGy and 40 kGy, respectively. The number of passes of wire or cable through the beam has its limitations, since
sufficient spacing is required between passes to avoid overlap or shadows.
The overall energy utilization coefficient can vary anywhere between 0.20
and 0.80.[46] Leading manufacturers of EB equipment provide formulas for
the calculation of required energy[47] and processing rate for the given dimensions of a wire and cable.[47–50]

9.1.2 Materials

Depending upon the end-use requirement, wire jacketing is most often made from formulated polyethylene. Blends of polyethylene and ethylene-propylene rubber are used if greater flexibility is needed, especially as the diameter of the jacketing increases as for cables. Another possibility is EPDM rubber with ENB monomer. Table 9.4 presents a typical radiation cross-linkable jacketing formulation.[51]

Hydrated aluminum oxide is a preferred flame retardant that liberates its water of hydration when exposed to flames, in contrast to chlorinated materials, which give off toxic gases as by-products. The paraffin oil is a processing aid that enhances the ability to extrude such materials. The silane is a coupling agent that improves the interaction between the polymers and the aluminum trihydrate. Trimethylolpropane trimethacrylate (TMPT) enhances the radiation response.

When enhanced temperature resistance is required, polyvinylidene fluoride or other fluoropolymers are used.[52] Fluoropolymers have the advantage of being oil resistant and flame retardant but are also more expensive base materials. PVDF is one of the materials that is very easy to cross-link by EB radiation.

9.1.3 Recent Developments and Trends

Radiation cross-linking of wire and cable insulations has been growing steadily over the past 33 years. On the weight bases, more wire insulation than cable insulation has been cross-linked by electron beams, with most of the wire being used in electronics. Recently, most of the growth was in flame-retardant

TABLE 9.4

Example of an Electron Beam-Curable Fire-Resistant Insulating Material

Material	Amount (phr[a])
Vistalon EPDM with ENB[b]	100
Hydrated aluminum oxide	100
Paraffinic process oil	50
Silane A 172	2
Antioxidant	1
TMPT[c] (75%)	3
Total	256

Source: ExxonMobil, Vulcanization of Vistalon Polymers, 2002.
[a] Parts per hundred parts of rubber.
[b] Ethylene norbornene (curing site in the elastomer).
[c] TMPT: Trimethylol propane trimethacrylate (prorad).

insulations. Besides PE and PVC, chlorinated polyethylene and chlorosulfonated polyethylene (Hypalon®) have been used.[53] Polypropylene (PP), similar to PE in price but stronger, can also be cross-linked by electron beams but requires prorads, such as trimethylolpropane triacrylate. These reduce the dose required for the cross-linking of PP considerably.[54] Another unique radiation cross-linked insulation is polymeric foam wire and cable insulation.[55]

9.2 Electron Beam Process in Tire Technology

A pneumatic tire is composed of several components, which must retain their shape and dimensions during the process of being built and during final vulcanization. Such components are the inner liner, chafer strip, white sidewall, veneer strip, and body and tread ply skims or skim coats (see Figure 9.6).[56] When these various parts of a tire are partially cured by radiation, they will not thin out or become displaced during the construction of the tire, nor will they thin out and flow during the vulcanization in the mold, as would components that were not irradiated. In some cases, material reduction is possible, and more synthetic rubber can be used in place of the more expensive natural rubber without a loss of strength.[56]

Body plies of a tire are various rubber-coated cord fabrics (see Figure 9.7). These are subjected to stresses when the tire is built. If a body ply does

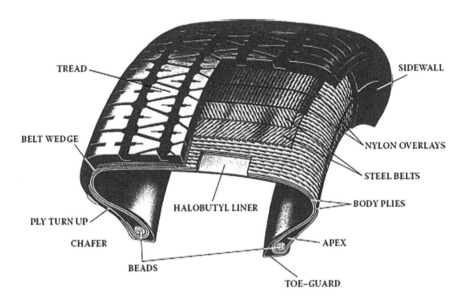

FIGURE 9.6
Cross section and components of a radial pneumatic tire. (Courtesy of B.F. Goodrich Co.)

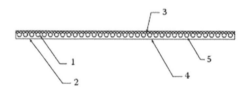

FIGURE 9.7
Cross section of a tire body ply with controlled depth of partial cross-linking. Legend: 1—Reinforcing cord fabric, 2—Skim coat, 3—Partially cross-linked skim coat by electron beam, 4—Building tack preserved, 5—Bleed cord.

not have sufficient green (i.e., uncured) strength, the cords will push through the rubber coating (skim) during the building and vulcanization of the tire, resulting in irregular cord placement or defect in the tire. For example, damage can occur from turning the ply around the bead during the building of the tire when the skim compound does not have adequate green strength. Green strength may be defined as that level of cohesive strength that allows an essentially uncross-linked polymer-based composition to deform uniformly, under stress, without sagging or nonuniform thinning (necking).

Tires are built most commonly on expandable drums. When the drum is expanded to shape the green (i.e., uncured) tire, the cord spacing in the body plies must be maintained. In tread plies, correct cord placement must be maintained during the change in their angle occurring during the shaping in the vulcanization process. Thus, the tread ply skim compound, like the body skim compound, has to have adequate green strength. Another important property of the rubber material in the tire building is the building tack, which is essentially adhesion to itself (autohesion). The level of building tack is reduced with the degree of cross-linking of the rubber compound. Thus, the green strength and building tack must be in good balance.

The construction of a tubeless tire starts with placing the inner liner on the building drum. Then the different plies and other components are added, including the tread, which is applied last. Then the tire is generally expanded into torroidal shape by air pressure. Additional expansion occurs in the heated curing mold during vulcanization of the tire. During the expansion of the green tire and vulcanization, the inner liner tends to decrease in thickness (thins out) and flow considerably, especially in the shoulder area, depending on the green strength of the inner liner compound. Since the inner liner has the function to maintain the inflation pressure of a tire during its service, its gauge is critical to the air permeation. Thus, maintaining its thickness in the curing process is very important. Normally, this is ensured by using a thicker inner liner than necessary. Partial cross-linking of the compound by irradiation makes this unnecessary, resulting in saving material and cost.

The degree of partial cure has to be precisely controlled, as pointed out earlier, in order to ensure sufficient green strength and sufficient building tack necessary for the construction of the tire. The electron beam process is ideally suited for that. The partial cure or precure is done by simply placing the tire components on a conveyor belt, which is passed under the electron beam source, and exposing them to the proper irradiation dose. The dose depends on the particular elastomer (or elastomer blends) and type and amount of fillers, oils, antioxidants, and other compounding ingredients that more or less affect the radiation response of the compound. Typically, a dose of 0.1–20.0 Mrad (1–200 kGy) is required.[56]

The effects of electron beam irradiation on the rubber compounds for tires can be readily evaluated by measuring green strength and recovery. Green strength is measured in a fashion similar to that of tensile strength: straight-sided specimens typically 12.5 mm (0.5 in.) wide and 2.5 mm (0.1 in.) thick are stretched by a tensile tester (e.g., Instron), and the peak stress or tensile strength is recorded. The laboratory test results from an uncured sheet can show whether the material has sufficient green strength for the given tire component if the slope of its tensile stress-strain curve remains positive beyond the maximum extension to which the component will be subjected before it is vulcanized.[56] For example, the green strength of the inner liner compound for a radial tire can be shown to be sufficient if the slope of its tensile stress-strain curve is positive throughout the 100%–300% elongation range, because maximum extension of the inner liner during the forming of the tire falls into the lower end of this range. If extension in the crown section of the tire were to exceed the uniform deformation capability of the liner, nonuniform thinning would occur.[57]

The recovery may be evaluated by the conventional Williams plasticity test (ASTM D926), Procedure 2.2.1, at 100°C (212°F). Other tests may be used, although the ones mentioned here were found to correlate best with the behavior of the rubber compound and tire components during the building and vulcanization of the tire. Results from green strength and recovery tests from different compounds reported in Hunt and Alliger[56] are shown in Table 9.5, and green strength of irradiated ply skim stocks with different compositions are shown in Table 9.6. Physical properties from vulcanized bromobutyl inner liner compound (unaged and aged) are shown in Table 9.7.[56]

The irradiated components can be placed adjacent to sulfur-containing unvulcanized elastomeric compounds and, when necessary, coated with suitable adhesive between the contacting surfaces to ensure good bond after final cure of the tire.[56]

The equipment and technology have developed significantly over the past two decades the high-energy electron beam process that has been in use. The original equipment was shielded by concrete vaults, which were 70–80 ft long, 35–40 ft high, and 16–20 ft wide. The walls of the treatment area might have been 4–6 ft thick. Pay-off and take-up equipment would add another 60–80 ft to the overall length of the system. The tire parts were treated almost exclusively off-line.[55–57] Due to a strong push by tire manufacturers to make

TABLE 9.5

Effect of Irradiation on Green Strength and Recovery of Different Tire Compounds

Compound	Property	Dose, kGy				
		0	50	100	150	200
Innerliner	Green strength, lbs[a]	3.0	15.3	20.0	22.9	33.8
	Recovery, %[b]	8.0	31.5	42.5	80.0	—
Chafer strip	Green strength, lbs[a]	22.8	35.1	42.1	52.0	58.2
	Recovery, %[b]	9.0	39.0	48.0	54.0	—
White sidewall	Green strength, lbs[a]	8.2	17.6	20.0	21.2	—
	Recovery, %[b]	10.0	35.5	113.5	110.0	—
Veneer strip	Green strength, lbs[a]	13.0	17.5	21.7	26.0	—
	Recovery, %[b]	10.5	35.5	71.0	118.5	—

[a] Peak value in tensile test.
[b] Plasticity. Williams (ASTM D928, Procedure 2.2.1).

TABLE 9.6

Green Strength of Irradiated Ply Skim Stocks with Different Elastomer Compositions, (lbs)

Type of Skim Stock	Nominal Radiation Dose, kGy				
	0	10	20	30	50
NR/ Solution SBR 50/50	3.2	4.6	6.6	8.5	11.3
100% Solution SBR	0.6	0.8	1.7	3.2	—
SBR/BR 75/25	1.5	—	—	—	25.7
100% BR	3.5	8.5	13.6	18.2	—
100% SBR	0.4	2.0	5.8	9.4	—

TABLE 9.7

Physical Properties from Vulcanized Bromobutyl Inner Liner Compound

Property	Nominal Radiation Dose, kGy				
	0	10	20	30	50
(a) Unaged					
Hardness, Durometer A	53	50	48	46	46
100% Modulus, MPa	1.2	1.1	1.0	0.8	0.9
300% Modulus, MPa	4.8	4.3	4.3	3.6	4.0
Tensile strength, MPa	9.9	8.7	8.0	6.9	7.5
Elongation at break, %	640	600	610	670	620
Peel adhesion at RT, kN/m	4.2	3.2	6.4	3.8	3.6

(*Continued*)

TABLE 9.7 (*Continued*)

Physical Properties from Vulcanized Bromobutyl Inner Liner Compound

	Nominal Radiation Dose, kGy				
Property	0	10	20	30	50
Peel adhesion at 100°C, kN/m	6.3	3.7	3.9	1.1	1.4
Fatigue, kcy (cured to rheometer optimum)	130	165	160	155	110
(b) Aged 240 h at 125°C					
Hardness, Durometer A	60	53	55	54	55
100% Modulus, MPa	2.2	1.8	1.8	1.8	1.9
300% Modulus, MPa	6.9	6.4	6.4	6.3	6.4
Tensile strength, MPa	8.4	7.5	7.5	7.2	7.3
Elongation at break, %	420	390	400	390	380
Fatigue, kcy (cured to rheometer optimum)	25	60	50	45	40

Source: Hunt, E.J. and Alliger, G., *Radiat. Phys. Chem.*, 14, 39, 1979.

the equipment more user-friendly, the accelerators were redesigned to produce radiation in the forward direction, which required radiation shielding only in the front and on sides.[56] Furthermore, the vault was placed below the ground level, and consequently, the factory floor provides the bulk of the shielding. Further improvements were in designing the shielding structure from a combination of lead and steel, replacing the traditional concrete vault.

The components that are supported by cords or fabric can be treated online easily by using a system of pulleys. This, unfortunately, is not yet possible for unsupported components, such as inner liner and other low-tension materials.[58,59]

The current technology is very cost-effective. If applied correctly in line with other processes, such as extrusion or calendering, it can produce substantial savings by the reduction in material or the tire component's weight. The reduction of overall thickness can be as much as 20%.[59]

9.3 Electron Beam Process in the Manufacture of Polyolefin Foams

The thin, highly stretched expanding foam cells of thermoplastic melts are unstable and bound to rupture at elevated temperatures when not stabilized. This is particularly important when chemical blowing agents are used, since they require relatively high temperatures to decompose. One of the methods of stabilization of polyolefin melt is cross-linking. Cross-linking not only stabilizes bubbles during expansion but also enhances the cellular product from thermal collapse, which is necessary in some applications.

Electron beam irradiation is one of the methods of cross-linking in this process.[60] The other methods use peroxide,[61,62] multifunctional azide,[63–66] or an organofunctional silane.[67] Polyethylene resins respond to electron beam irradiation well because the rate of cross-linking significantly exceeds the chain scission. Polypropylene (PP) is prone to β-cleavage, which makes it difficult to cross-link by a free radical process.[67] For that reason, PP resins require cross-link promoters, such as vinyl monomers, divinyl benzene,[68,69] acrylates and methacrylates of polyols,[70,71] polybutadiene, and others. These compounds promote cross-linking of a polyolefin polymer through inhibiting the decomposition of polymer radicals,[72] grafting,[73] and addition of polymerization-type reactions.[67]

Optimum cross-linking is the most critical requirement for optimum foam expansion. The *gel level* or *gel fraction* is defined as the fraction not soluble in boiling xylene.[74] Excessive cross-linking restricts foam expansion, while insufficient cross-linking results in bubble rupture.[75] The window of optimum cross-linking is fairly narrow; the gel level at the inception of gel expansion needs to be about 20%–40%, preferably 30%–40%.[76,77] An LDPE resin with a lower value of melt index (higher molecular weight) and a lower density becomes more efficiently cross-linked. A lower-density LDPE has more chain branching and therefore more tertiary hydrogens, which provide cross-linking sites. Other factors influencing cross-linking are molecular weight distribution and long-chain branching.[76–78] For typical polyethylene, a dosage ranging from 10 to 50 kGy is sufficient to attain the final gel level of 30%–40%.

The selection of EB equipment depends on the foam thickness and production rate. The foam manufacturers throughout the world are using equipment with electron accelerators with voltages ranging from 0.5 to 4 MV and power ratings from 10 to 50 kW. The penetration depth of a 1 MV unit is approximately 3 mm (0.12 in.). Irradiation on both sides doubles the thickness capability.[79]

9.3.1 Foam Expansion and Its Control

The cross-linked sheet is expanded by means of a chemical blowing agent such as azodicarbonamide, which decomposes in the temperature range between 200°C and 210°C (392°F–410°F). In the radiation cross-linking process,[80,81] the cross-linking reaction is completed before the decomposition of the blowing agent and foam expansion. Since the decomposition of the blowing agent and foam expansion occur simultaneously, this process can run twice as fast as peroxide cure because the heating rate is not restricted by the cross-linking rate.[76] The cell size is easier to control in the cross-linked sheet, since the cross-links provide the restricting force against cell growth. Thus, radiation cross-linking favors cell nucleation, and fine uniform cells (typically 0.2–0.3 mm in size) are easily achieved.[82]

9.3.2 Manufacturing Processes

A cross-linked polyolefin foam sheet is produced by two methods using chemical cross-linking and by two methods of radiation cross-linking. The two well-established manufacturing processes for polyolefin foams using radiation cross-linking are the *Sekisui* process and *Toray* process. The differences between these two manufacturing methods are mainly in the expansion step, which is almost always done separately. However, the production of the foamable sheet and the cross-linking step are similar. The first step is a uniform mixing of the blowing agent into the polymer melt, followed by sheet extrusion in an extrusion line. Then the cross-linking to the desired degree follows. The flow diagram for this process is shown in Figure 9.8.

The Sekisui process employs a vertical hot air oven. The foamable sheet is first preheated in the preheating chamber by infrared heaters to about 150°C (302°F) while being supported by an endless belt. The preheated sheet is then expanded in the foaming chamber to temperatures exceeding 200°C (392°F). The expanding sheet supports itself by gravity in the vertical direction. A special tentering device is used to prevent the development of wrinkles in the transverse direction.[83] The advantages of the Sekisui process are the capability of manufacturing a thin sheet and the low-energy consumption inherent in the vertical oven.[84] A schematic of the Sekisui foaming oven is shown in Figure 9.9.

In the Toray process, the foamable sheet is expanded as it floats on the surface of molten salts and is heated from the top by infrared lamps. The molten salt mixture consists of potassium nitrate, sodium nitrate, and sodium nitrite.[85] The salt residues from the surface of the foam are blown off by hot air and stripped in water. The Toray process is suitable to produce cross-linked PP foam sheet as well as polyethylene foam sheet. In fact, Toray was the first one to produce commercial PP foam.[86]

9.3.3 Comparison of Chemical and Radiation Processes

Both processes have advantages and disadvantages. Currently, they have about an equal share of the global market. The initial investment cost in the irradiation equipment is high, but it is considerably more productive than

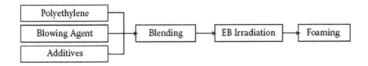

FIGURE 9.8
Flow diagram of the manufacturing process for polyolefin foams using radiation cross-linking.

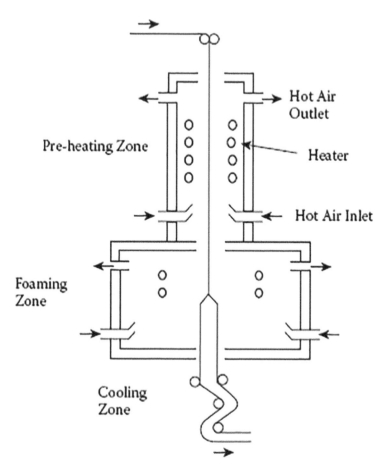

FIGURE 9.9
Schematic of Sekisui foaming oven for polyolefin foams.

chemical cross-linking, having the additional advantage of uniform product quality and flexibility of feedstock selection. The product from radiation
cross-linked polyolefins is thin, with fine cells and smooth white surfaces.

Correctly foamed polyolefins consist of small regular closed cells. Because
of this, they have a high heat and sound insulation capacity and excellent
shock-absorbing capability. They are used for the insulation of central heating
pipes, in the packaging industry, and in sports and leisure articles for protection of the head, knee, shin, and elbow. In the health-care industry, polyolefin
foams are used as backing for medical devices. In automotive applications,
polyolefin foams are used for safety and protection in dashboards and door
panels (Figure 9.10), and most notably as cushioning under the interior header.

The comparison between chemical and radiation processes is shown in
Table 9.8.[77,87]

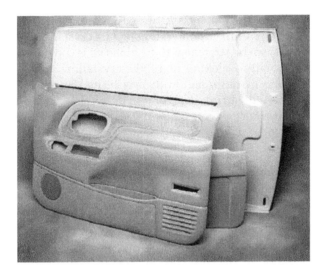

FIGURE 9.10
Crosslinked PE closed cell in automotive application. (Courtesy A.J. Berejka.)

TABLE 9.8

Comparison of Radiation and Chemical Cross-Linking Processes for Producing Polyolefin Foams

Item	Radiation Cross-Linking	Chemical Cross-Linking
Process control	Easy	Difficult
Production rate	High	Slow
Equipment	Cumbersome	Simple
Cost	Decreases with production volume	Relatively constant
Blowing agent selection	Easier	More difficult
Product thickness, mm	3–6	5–16
Typical cell size, mm	0.2–0.4	0.5–0.8
Cross-link level, %	30–40	60–70

9.4 Electron Beam Process in the Production of Heat-Shrinkable Materials

Polyolefins, especially polyethylene, can be cross-linked into a material that is elastic when heated. The structure of polyolefins, normally entangled long chains, includes crystalline and amorphous regions. Upon heating above the crystalline melting point of the polymer the crystalline regions disappear.

When cross-linking the polymer, a three-dimensional network is formed. After heating the cross-linked material above its crystalline melting point, the elastic network can be stretched. If the material is cooled in the stretched state below its crystalline melting point, the crystalline regions reform and the material retains its deformation. If it is heated again above the melting temperature, it will shrink back to its original state. This phenomenon is referred to as memory effect and is used for heat-shrink tubing and in packaging.

9.4.1 Heat-Shrinkable Tubing

Heat-shrinkable tubing is made typically from polyolefins, PVC, polyvinyl fluoride, PTFE, their blends, or blends with other plastics and elastomers. The formulations may be designed for chemical resistance, heat resistance, flame resistance, etc.[88]

There are essentially two methods to produce shrink tubing from an extruded tube. One of them involves expanding the tube, which had been cross-linked by radiation, heated to temperatures above its crystalline melting point into a sizing tool. The diameter of the sizing tool determines the amount of stretch. The stretched tubing is then cooled down to temperatures below the crystalline melting point in the sizing tool, thus "freezing in" the strain. This method is described in the original Raychem patent (see Figure 9.11). The patent specifies the irradiation dose to be at least about 2×10^6 rad (20 kGy).[89] Current practice is to irradiate polyethylene tubing with electrons at 1.0–2.0 MeV and at a dose of 200–300 kGy. Excessive doses should not be applied; otherwise, the material will have a poorer aging resistance.[90]

The second method is designed to expand the tubing heated above its crystalline melting point into a stationary forming tube or mold by air pressure. The tubing is surrounded by a moving tape over the entire length of the forming device. The tape is formed from a material that does not adhere to the cooled plastic, so it can be readily stripped from the expanded tubing as it emerges from the forming device. The arrangement of the patented process[91] is shown in Figure 9.12.

Figure 9.13 depicts the process of producing heat-shrinkable polyethylene tubing and the use of the tubing to cover a wire joint. In step I, the tubing is extruded and irradiated to obtain a gel fraction of 40%. In step II, the irradiated tubing is heated up to 140°C (285°F) and expanded by the use of vacuum or pressure to about twice its original diameter, holding its length constant. Steps III and IV show the use of heat-shrinkable tubing to cover a joint of two wires. The tubing is centered over the joint and heated by a hot air gun to a temperature above the crystalline melting point of polyethylene. The sleeve shrinks and becomes a form-fitting cover for the joint.[92]

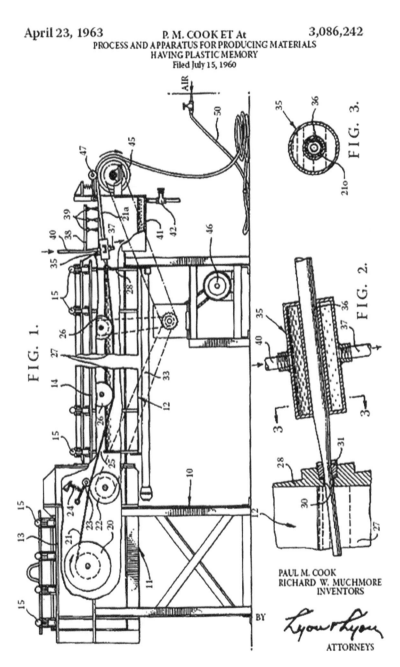

April 23, 1963   P. M. COOK ET At   3,086,242

PROCESS AND APPARATUS FOR PRODUCING MATERIALS
HAVING PLASTIC MEMORY
Filed July 15, 1960

PAUL M. COOK
RICHARD W. MUCHMORE
INVENTORS

BY

ATTORNEYS

FIGURE 9.11
Original Raychem patent for heat shrinkable tubing.

Jan. 3, 1967   R. TIMMERMAN   3,296,344

METHOD AND APPARATUS FOR EXPANDING PLASTIC TUBING

Filed May 29, 1963   2 Sheets-Sheet 1

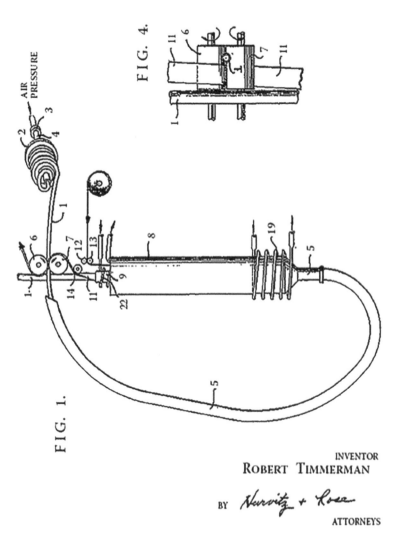

INVENTOR

ROBERT TIMMERMAN

BY *Hurvitz + Rose*

ATTORNEYS

FIGURE 9.12
Alternative patented method of producing heat shrinkable tubing.

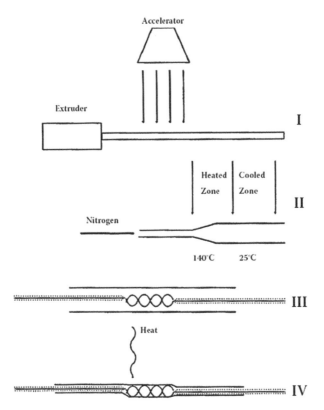

FIGURE 9.13
Production and application of heat-shrinkable tubing on a wire splice. (From *Radiation Processing of Polymers*, Singh, A. and Silverman, J., Eds., Carl Hanser Publishers, Munich, p. 20. With permission.)

Thin-walled (approximately 0.010–0.040 in. or 0.25–1.0 mm) tubing is usually selected by continuous service temperature, which may range from 90°C (194°F) for some PVC tubing up to 250°C (482°F) for tubing made from PTFE. The main use of thin shrink tubing is in electrical and electronic applications.

Tubing with thicker walls (typically in the range of 0.080–0.170 in. or 2–4 mm) is fabricated mainly from polyolefins and is used to cover splices in the telecom, CATV, and electric power industries. Often such tubing is combined with mastic or hot melt that aids in forming an environmental barrier for the splice. Diameters of the heavy wall tubing may be up to 7 in. (178 mm) or even 12–24 in. (300–600 mm) when used as corrosion protection sleeves on weld joints of gas and oil pipelines.[93]

9.4.2 Heat-Shrinkable Sheets and Films

Heat-shrinkable sheets (thickness 0.040–0.120 in. or 1–3 mm) and films (thickness 0.001–0.020 in. or 0.025–0.5 mm) are fabricated from many of the same materials as shrinkable tubing.[92] They are produced by extrusion as a tube, sheet, or blown film. Irradiation is done by the equipment as shown in Figure 9.14. Orientation (stretching) after irradiation can be performed by several methods, namely, by differentially heated and driven rolls (in the machine direction) or by a tenter frame (across the machine direction), as shown in Figure 9.15.

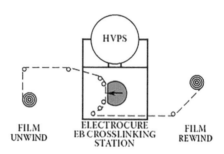

FIGURE 9.14
Electron beam processor for film conversion. (Courtesy of Energy Sciences, Inc.) (HVPS: high-voltage power supply. The arrow in the EB curing unit indicates the direction of the electron beam).

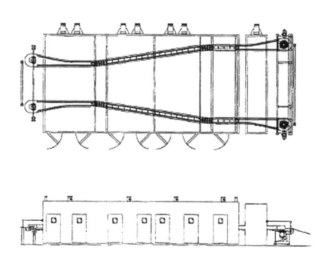

FIGURE 9.15
Tenter frame for transverse direction orientation of sheets and films. (Courtesy of Marshall & Williams Plastics.)

FIGURE 9.16
Production line for heat shrinkable films.

Heat shrink sheets are used for many of the same applications as heavy wall shrink tubing. The advantage of the sheet over a tube is that it can be conveniently slipped over the area to be protected, for example, over continuously installed cable. There are many methods to close the heat-shrinkable sheets, such as zippers, rail and channel, heat-sealable bonds, etc., some of them patented.[94–97]

Heat-shrinkable films have found wide use in food wrap and packaging. The original development was done in the late 1950s by W. R. Grace and Co., and the process is still used for much more sophisticated multilayer laminated films with superior barrier properties,[97–100] exceptionally high tear strength,[101] or for multilayer films with low shrink force for packaging of easily deformable articles.[102] The currently used doses for PE films are 200–300 kGy at electron energy in the range of 0.5–1.0 MeV.[89] A line for 10 EB units for heat-shrinkable films is shown in Figure 9.16.

9.5 Electron Beam Process in Coatings, Adhesives, Paints, and Printing Inks

These applications are a domain that UV and EB processes share, since both can directly convert reactive liquids into solids almost instantly. There are specific areas where EB irradiation is more suitable than the UV curing process. In general, these include applications where thick layers of coatings or adhesives are applied. Other instances are coatings with high levels of inorganic pigments or fillers, which usually cannot be cured by UV radiation because of their opacity. As pointed out earlier, the capital cost of standard EB curing equipment is considerably higher than that of a UV curing line. However, a recent trend is to build smaller electron beam processors operating at a much lower voltage (see Section 4.2.3). Such machines are considerably less expensive and consequently represent formidable competition to UV curing equipment in an increasing number of applications. At any rate, EB curing lines operate at much higher line speeds and compare favorably if they are used for continuous long runs. In this section, discussion will be limited to applications specific to the EB curing process or possibly a combination of EB and UV.

9.5.1 Magnetic Media

Magnetic media such as magnetic tapes for audio and video recording, program cards of various types, and computer diskettes have an important place in the electronic industry. The magnetic particles, such as gamma ferric oxide, ferric oxide with Cr^{+2}, chromium oxide, and iron-cobalt alloys, are dispersed in a solution of a binder.[103] The binders include polyurethanes, vinylchloride/vinyl acetate copolymers, vinylchloride/vinyl acetate/vinyl alcohol copolymers, polyacrylates, nitrocellulose, and others. Particularly important are blends of polyurethanes with other polymers.[104] The substrates used for tapes and other forms of magnetic media are polyvinylchloride, poly(ethylene terephtalate), and other film-forming polymers, with poly(ethylene terephtalate) currently being most widely used.[105]

The electron beam is very well-suited for the production of most of the magnetic media, such as high-density floppy disks and magnetic tapes. In this process, the substrate is coated with a coating containing a large volume of magnetic particles. The uncured coating passes first through a magnetic field to orient the particles and then through the EB processor. The high loading of the magnetic particles excludes the use of UV cure. The electron beam offers several advantages: The high reaction rate virtually eliminates surface contamination closely associated with thermal cure. The high degree of cross-linking of the binders used in the coatings gives a tougher and more wear-resistant surface.

9.5.2 Coatings

Most of the EB units sold in the recent past were mainly for release coating applications. New chemistry has been commercialized, but further developments are needed to satisfy this market segment.[106]

Another process where electron beam equipment is widely used is curing of overprint varnishes (OPVs). OPV is applied over a printed surface to protect the ink layer and to improve the appearance of the product.[106] The exceptionally high gloss of such finishing coatings and varnishes is achieved specifically by EB cure.[107,108]

Typical applications of EB curing in protective and decorative coatings include[109]:

- Wood finishes (doors, front panels of furniture, lamination and lacquer finishes)
- Paper coatings and finishes (paper laminates for wood decoration, high-gloss paper for gift wrapping, lacquer on metallized paper)
- Film coatings (antistatic coatings, antiscratch coating on phone and other cards)

Coatings on *rigid substrates*, mainly on wood (doors, furniture, laminated panels), often have a fairly thick coating (up to 200 g/cm²) and consequently require electrons with relatively high energy. However, production rates are mostly dependent on other operations, such as feeding, discharge, and polishing, and therefore, low to moderate beam powers are sufficient. An example of curing the top lacquer on doors is shown in Figure 9.17.

Coatings on *flexible substrates* are considerably thinner, typically in the range of 130 g/cm², but the production rates are high (100–300 m/min); therefore, lower energy electrons and moderate to high power beams are required.[109]

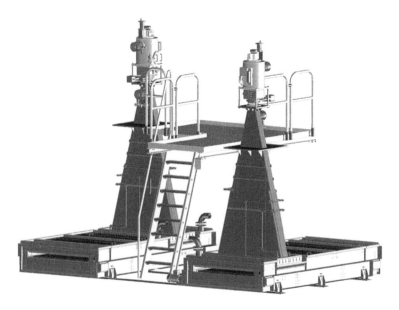

FIGURE 9.17
Curing of top lacquer on doors. (Courtesy of Elektron Crosslinking AB.)

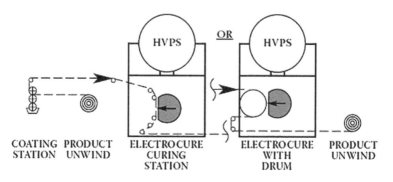

FIGURE 9.18
Electron beam processor for curing coatings. (Courtesy Energy Sciences, Inc.) (HVPS: high-voltage power supply. The arrow in the EB curing unit indicates the direction of the electron beam).

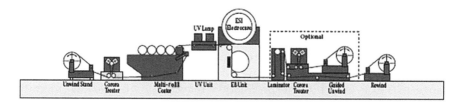

FIGURE 9.19
Coating line for a hybrid UV/EB curing process for coatings. (Courtesy of Faustel Corporation.)

In some cases, a hybrid system (a combination of UV and EB cures) is used to ensure a sufficient degree of cross-linking through the coating or a sufficient adhesion to the substrate. An additional benefit of such a hybrid system is the control of gloss and surface finish.[110]

An example of EB curing of coatings is shown in Figure 9.18, and that of a curing line for a hybrid UV/EB process is shown in Figure 9.19.

9.5.3 Printing and Graphic Arts

With the advent of less expensive and more compact electron beam processors, a growing number of printers/converters in graphic arts are exploring the possibility to use this process as an alternative to conventional ink/coating drying.[111]

Currently, the following EB cure applications in printing and graphic arts are reported in the literature[108,112]:

- High-gloss cosmetic and cigarette packaging
- Greeting cards

- Aseptic packaging
- Record album jackets
- Stamps
- Banknotes

The most common installation of EB equipment is at the end of a web offset press used for the production of folding cartons (Figure 9.20). Offset (lithographic) printing uses paste inks, which are designed to be "wet trapped" without any interstation drying. That allows placing a single EB unit at the end of the press.[113] The development of compact low-voltage EB equipment has allowed for their use to cure inks on flexible packaging, polyester, and polypropylene films and labels.[114,115] The low-voltage permanent vacuum modular equipment has potential to be used for EB curing between the stations.[113]

9.5.4 Adhesives

In the field of adhesives, electron beam curing has mainly been used in pressure-sensitive and laminating adhesives.[116]

9.5.4.1 Pressure-Sensitive Adhesives (PSAs)

The main property that distinguishes a pressure-sensitive adhesive from other types of adhesives is that it exhibits a permanent and controlled tack. This tackiness is what causes the adhesive to adhere instantly when it is pressed against a substrate. After it adheres, the PSA should exhibit tack, peel, and shear properties, which are reproducible within narrow limits. This requires that the adhesive layer be only slightly cross-linked.[117] PSAs are based on polymers with low T_g, typically in the range of $-74^\circ C$–$+13^\circ C$.[118]

FIGURE 9.20
Low-voltage EB equipment on web offset press. (Courtesy of PCT Ebeam and Integration LLC.)

Because of the precise control of the degree of cross-linking attainable by the electron beam process, it is well suited for the production of PSAs. Recent developments in EB design, particularly lower voltage, and materials developments have made electron beam curing of pressure-sensitive adhesives affordable, practical, and possible.[119–121]

9.5.4.2 Laminating Adhesives

Laminating adhesives are used to bond layers of different materials together. There are many possible combinations, including different polymers, copolymers, papers, foils, fabrics, films, metals, glass, etc. The bond between the individual layers has to be sufficiently strong to hold the laminate together. This is accomplished by applying the adhesive in the molten state (hot melt) and creating the bond upon cooling or by cross-linking. Cross-linking can be accomplished by using a reactive adhesive, consisting of two components, which react when mixed (e.g., polyurethanes or epoxies), or by radiation.

Cross-linking increases the cohesive strength of the adhesive and consequently the bond strength. Many of the materials used for laminations are transparent to light, and these can easily be cured by UV or visible light. However, if the laminated materials are opaque to UV and visible light, they may be cross-linked by electron beam because high-energy electrons penetrate paper, foils, and fabrics.

Electron beam curing has been used for curing laminating adhesives in flexible packaging, and this application has been growing rapidly after the introduction of low-voltage compact EB processors. The advantages of EB curing of laminating adhesives are[122,123]:

- Adhesives do not use solvents.
- Adhesive is one part chemistry (no mixing needed).
- The shelf life is long (more than 6 months).
- Adhesive remains unchanged until it is cured.
- No multiroll coating is needed.
- No complex tension controls are needed.
- Adhesive bond is established almost instantly.
- Real-time quality control exists.
- In-line processing and immediate shipment are possible.
- There is easy cleanup.
  Advantages over UV curable laminating adhesives are[123]:
- EB penetrates opaque films.
- EB produces typically higher conversion.
- Most EB systems cure without photoinitiators (simpler formulation, cost savings, fewer residues and extractables).

Recently, a solvent-free dual-cure laminating adhesive based on the combination of polyurethane chemistry and EB cure was developed that exhibits improved properties compared to laminates prepared by either method alone.[124]

9.5.5 Polymeric Fiber-Reinforced Composites

Fiber-reinforced composites are materials combining high strength with a low weight. Polymeric composite materials consist of a thermoset or thermoplastic matrix and reinforcing fibers. The main functions of the matrix are to act as a binder for the fibers, to transfer forces from one fiber to the other, and to protect them from environmental effects and effects of handling. The reinforcing fibers can be continuous or discontinuous (short fibers) and may be oriented within the matrix at different angles. Short fibers are frequently dispersed in the matrix randomly, although some composite materials may contain oriented short fibers.

The performance characteristics of a composite material depend on the type of reinforcing fiber (its strength and stiffness), its length, fiber volume fraction in the matrix, and the strength of the fiber-matrix interface. The presence of voids and the nature of the matrix are additional but minor factors.

The majority of commercially available polymeric composites are reinforced by glass fibers, carbon fibers, aramid fibers (e.g., Kevlar) and, to a lesser degree, boron fibers. In some cases, hybrid composites are made that contain some combinations of fibers.

Matrix materials for commercial composites are mainly liquid thermosetting resins, such as polyesters, vinyl esters, epoxy resins, and bismaleimide resins. Thermoplastic composites are made from polyamides, polyetheretherketone (PEEK), polyphenylene sulfide (PPS), polysulfone, polyetherimide (PEI), and polyamide-imide (PAI).

Thermosetting composites are cured at either ambient or elevated temperatures to obtain a hard solid by cross-linking. The use of radiation cross-linking decreases the cure time considerably. In particular, an electron beam has been used successfully in many instances. For example, glass-fiber-reinforced composites cured by electron beam have been used for the production of cladding panels.[125]

The production of materials having good mechanical properties matching those produced by conventional thermal methods has been achieved by using EB processing, and a growth in these applications is expected.[124,125] Graphite-fiber-reinforced composites with low stress and exhibiting little shrinkage upon cure have been produced with a performance comparable to that of state-of-the-art toughened epoxy resins.[126–128] A layer-by-layer electron beam curing process of filament-wound composite materials using low-energy electron beams has been developed. This process has a wide range of applications in the aerospace technology.[129,130] A variety of different curing methods leading to successful products are described in the literature.[131–134]

FIGURE 9.21
Composite missile body irradiated by electron beam. (Courtesy A.J. Berejka.)

The article by Berejka[135] covers the state of the art in electron beam technology, material technology, and product-forming technology as applied to structural carbon-fiber-reinforced polymeric composites. An example of a composite missile body is shown in Figure 9.21.

A great deal of work was done to develop methods for using electron beam technology for the manufacture and repair of advanced composites. Many benefits have been identified for EB curing fiber-reinforced composites, including lower residual stresses that result from curing at ambient or subambient temperatures, shorter curing time for individual components (minutes vs. hours), improved material handling (the resins have unlimited shelf life), and possible process automation in the placement of fiber reinforcement. The only process EB cannot use is resin transfer molding (RTM), since the electron beam cannot penetrate massive molds used to mold the composite parts.[136] So far, mainly EB curable cationic-initiated epoxy resins have been used. Although most of the development work was geared toward aerospace, the technology is applicable to other industries, including automotive, consumer goods, etc.

The development work also yielded UV curable adhesives for aluminum to aluminum, graphite to aluminum, and graphite to graphite.

So far, several parts have been produced by applying EB technology:

- Lower- and upper-wing assembly
- Liquid hydrogen tank

Another viable application is the use of EB curing in the repair of aircraft composite parts. It was determined that most of the repairs are completed in 50%–70% of the time required for a normal thermal repair.[136]

9.5.6 Hydrogels

Hydrogels are three-dimensional hydrophilic polymeric structures able to absorb large quantities of water. They are synthesized by polymerization of hydrophilic monomers. The extent of the reversible swelling and deswelling property of these materials is known to depend on the nature of both inter-molecular and intramolecular cross-linking, as well as the degree of hydro-gen bonding in the polymer network. Increasingly, these hydrogels are being utilized in a variety of applications, including drug release, biosensors, tissue engineering, and pH sensors. To date, a variety of compounds have been utilized in the synthesis of hydrogels. Examples of monomers employed for this purpose include N-isopropylacrylamide (NIPAM), vinyl alcohol, ethylene glycol, and N-vinyl pyrrolidone (NVP).[137] These monomers can be combined with natural polymers, such as agar.[138]

Radiation techniques are very suitable tools for the production of hydrogels. An example of a mature technology of this kind is hydrogel wound dressings, now being produced on large scale.[139] Other practical applications are contact lenses (silicone hydrogels, polyacrylamides), dressings for the healing of burn victims and other hard-to-heal wounds, disposable diapers, and similar highly absorbing products.[138]

The most widely used hydrogels are based on polyethylene oxides (PEOs) dissolved at relatively low concentrations in water (typically 4%–5%). Modest radiation exposure (less than 10 kGy) is needed to form the gel. Gels as thick as 2 mm (80 mils) are being produced.

Current applications include a hydrogel-based system for anticancer therapy due to local drug delivery, systems for encapsulation of living cells, a new approach to the synthesis of polymeric material for intervertebral discs implant, temperature-sensitive membranes, hydrogel phantoms of a 3D radiation dosimeter for radiotherapy, degradation-resistant nanogels and microgels for biomedical purposes (e.g., synovial fluid substitute), and hydrogel-based dietary products.[139]

Another application reported[140] is preparing hydrogels from a mixture of polyvinyl alcohol (PVA) and carboxymethyl cellulose (CMC) using freezing and thawing and EB irradiation. The blend having the composition 80/20 (CMC/PVA) was used as a superabsorbent additive in the soil for agriculture. Moreover, the water retention is increased in the soil containing this hydrogel. Thus, this type of hydrogel can be used to increase water retention in desert regions.

Other hydrogels are prepared from poly(α-hydroxylic) acids, for example, lactic acid, polyacrylic acid, acrylamides and acrylated polyvinyl alcohol,[141] and polyethylene glycol.

9.5.7 Production of Fluoroadditives

Fluoroadditives, or PTFE micropowders, are finely divided low-molecular-weight polytetrafluoroethylene powder. In general, they consist of small

TABLE 9.9

Effect of Degradation on Properties of Polytetrafluoroethylene Resins

Resin	Molecular Weight	Typical Melt Viscosity, Pa.s
Molding resin	10^6	10^{10}
Fluoroadditive (degraded resin)	10^4–10^5	10–10^4

particles of the order of several micrometers. They are produced by several methods, namely degradation by heat, ionizing irradiation, and controlled polymerization. The raw material for the degradation methods is very often sintered or unsintered PTFE scrap, molding powders, production scrap and post-consumer PTFE articles.

The most economical way to produce fluoroadditives is a continuous process of irradiation of the resin and/or scrap under electron beam source.[142] The processed material is spread on a conveyor belt in a specific layer thickness and is exposed to the required dose. Usually, the dose for sufficient degradation is 500 kGy or more. The advantage of multiple passes is that the material is allowed to cool down, which prevents it from sticking together and causing problems in the next step. The material heats up during irradiation so the temperature is held at 121°C, which is sufficient to prevent sticking. The degradation of the PTFE resin generates off-gasses such as hydrofluoric acid that must be removed by ventilation of the processing area.[143]

The degradation process, i.e., reduction of molecular weight decreases rapidly with the absorbed dose. The effect of degradation of the PTFE resin is illustrated in Table 9.9.

When the irradiated resin has received sufficient dose, it is ground to the desired particle size. The most widely used grinding method is in fluid energy mills, more commonly known as jet mills, in which a compressed fluid such as highly compressed air is used as the source of energy.[143]

The ground micropowders have an apparent density of 400 g/L and a melting point of 321°C–327°C. Specific surface area ranges from 5 to 10 m^2 and the average particle size ranges between 3 and 12 µm.

PTFE micropowders are used as additives to lubricants (oils and greases), to plastics and elastomers to reduce friction and improve extrusion, to printing inks to reduce blocking and improve abrasion resistance, and to coatings to reduce wear and friction and to increase water and oil repellency.[143]

9.6 Other Applications for Electron Beam Radiation

Other applications are proven and effective EB processes but are often limited by the size of a given market or by still developing commercial acceptance. These are[144]:

- Battery separators
- Filter membranes
- Artificial joints

References

1. Mehnert, R., Bogl, K. W., Helle, N., and Schreiber, G. A., in *Ullmann's Encyclopedia of Industrial Chemistry*, Vol. A22 (Elvers, B., Hawkins, S., Russey, W., and Schulz, G., Eds.), VCH Verlagsgesselschaft, Weinheim, p. 484 (1993).
2. McGinnis, V. D., in *Encyclopedia of Polymer Science and Engineering*, Vol. 4 (Mark, H. F., and Kroschwitz, J. I., Eds.), John Wiley & Sons, New York, p. 445 (1986).
3. Yongxiang, F., and Zueteh, M., in *Radiation Processing of Polymers* (Singh, A., and Silverman, J., Eds.), Hanser, Munich, p. 72 (1992).
4. Levy, S., and DuBois, J. H., *Plastic Product Design Handbook*, Van Nostrand Reinhold, New York (1977).
5. Rosen, S., *Fundamental Principles of Polymeric Materials*, John Wiley & Sons, New York (1982).
6. Williams, H., *Polymer Engineering*, Elsevier, New York (1975).
7. Roberts, B. E., and Verne, S., *Plast. Rubber Process Appl.*, 4, p. 135 (1984).
8. Hochstrasse, U., *Wire Industry*, 46, January (1985).
9. Markovic, V., *IAEA Bulletin*, IAEA (International Atomic Energy Agency), Vienna, Austria (1985).
10. Brand, E. S., and Berejka, A. J., *World*, 179, p. 49 (1978).
11. Clelland, M. R., *Radiat. Phys. Chem.*, 18, p. 301 (1981).
12. Yongxiang, F., and Zueteh, M., in *Radiation Processing of Polymers* (Singh, A., and Silverman, J., Eds.), Hanser, Munich, p. 73 (1992).
13. Barlow, A., Hill, L. A., and Meeks, L. A., *Radiat. Phys. Chem.*, 14, p. 783 (1979).
14. Bly, J. H., *Radiat. Phys. Chem*, 9, p. 599 (1977).
15. Yongxiang, F., and Zueteh, M., in *Radiation Processing of Polymers* (Singh, A., and Silverman, J., Eds.), Hanser, Munich, p. 74 (1992).
16. Bruins, P. F., *Plastics for Electrical Insulation*, Wiley-Interscience, New York (1968).
17. Harrop, P., *Dielectrics*, Butterworth, London (1972).
18. Schmitz, E.V., in *Testing of Polymers*, Vol. 1, Chapter 1 (Schmitz, J. V., Ed.), Wiley-Interscience, New York (1968).
19. Association of Edison Illuminating Companies, AEIC 5–79, *Specifications, PE Power Cables* (5–69 kV) (1979).
20. ASTM D150, *AC Loss Characteristics and Dielectric Constants*, American Society of Testing Materials, Philadelphia, PA.
21. Blythe, A. R., *Electrical Properties of Polymers*, Cambridge University Press, London (1979).
22. Yongxiang, F., and Zueteh, M., in *Radiation Processing of Polymers* (Singh, A., and Silverman, J., Eds.), Carl Hanser Verlag, Munich, p. 76 (1992).
23. Wintle, H. J., in *Radiation Chemistry of Macromolecules* (Dole, M., Ed.), Academic Press, New York (1972).

24. Yongxiang, F., and Zueteh, M., in *Radiation Processing of Polymers* (Singh, A., and Silverman, J., Eds.), Carl Hanser Verlag, Munich, p. 77 (1992).
25. Charlesby, A., and Fydelor, P. J., *Radiat. Phys. Chem.*, 4, p. 107 (1972).
26. Schnabel, W., *Polymer Degradation*, Maxmillan, New York (1981).
27. Ueno, K., *Lecture Notes on Formulation of Radiation Cross-Linking Compounds*, IAEE Training Course, Shanghai, September (1989).
28. DuBois, J. H., and John, F. W., *Plastics*, Van Nostrand Reinhold, New York (1974).
29. Matsuoka, S., *IEEE Trans. Nuclear Sci.*, NS-23, p. 1447 (1976).
30. Sasaki, T., et al., *Radiat. Phys. Chem.*, 14, p. 821 (1979).
31. Barlow, A., Briggs, J. W., and Meeks, L. A., *Radiat. Phys. Chem.*, 18, p. 267 (1981).
32. Chapiro, A., *Radiation Chemistry of Polymer Systems*, Wiley-Interscience, New York (1962).
33. Salovey, R., and Gebauer, R., *J. Polym. Sci.*, A-1, p. 1533 (1972).
34. Yongxiang, F., and Zueteh, M., in *Radiation Processing of Polymers* (Singh, A., and Silverman, J., Eds.), Hanser, Munich, p. 80 (1992).
35. Waldron, R. H., Mc Rae, H. F., and Madison, J. D., *Radiat. Phys. Chem.*, 25, p. 843 (1988).
36. Peshkov, I. B., *Radiat. Phys. Chem.*, 22, p. 379 (1983).
37. Yongxiang, F., and Zueteh, M., in *Radiation Processing of Polymers* (Singh, A., and Silverman, J., Eds.), Hanser, Munich, p. 81 (1992).
38. Becker, R. C., Bly, J. H., Cleland, M. R., and Farrell, J. P., *Radiat. Phys. Chem.*, 14, p. 353 (1979).
39. Cleland, M. R., *Technical Information Series, TIS 77-12*, Radiation Dynamics, Melville, NY (1977).
40. McLaughlin, W. L., et al., *Dosimetry for Radiation Processing*, Taylor and Francis Group, London (1989).
41. Zagorski, Z. P., in *Radiation Processing of Polymers* (Singh, A., and Silverman, J., Eds.), Carl Hanser Verlag, Munich, Chapter 13 (1992).
42. Yongxiang, F., and Zueteh, M., in *Radiation Processing of Polymers* (Singh, A., and Silverman, J., Eds.), Carl Hanser Verlag, Munich, p. 82 (1992).
43. Studer, N., and Schmidt, C., *Wire J. Int.*, 17, p. 94 (1984).
44. Cleland, M. R., in *Radiation Processing of Polymers* (Singh, A., and Silverman, J., Eds.), Carl Hanser Verlag, Munich, p. 81 (1992).
45. Yongxiang, F., and Zueteh, M., in *Radiation Processing of Polymers* (Singh, A., and Silverman, J., Eds.), Hanser, Munich, p. 84 (1992).
46. Yongxiang, F., and Zueteh, M., in *Radiation Processing of Polymers* (Singh, A., and Silverman, J., Eds.), Carl Hanser Verlag, Munich, p. 81 (1992).
47. Luniewski, R. S., and Bly, J. H., in *Proceedings of Regional Technical Conference "Irradiation and Other Curing Techniques in the Wire Industry,"* Newton, MA, March 20–21 (1975).
48. Barlow, A., Biggs, J., and Maringer, M., *Radiat. Phys. Chem.*, 9, p. 685 (1977).
49. Clelland, M. R., *Accelerator Requirements for Electron Beam Processing*, Technical Paper TIS 76-6from Radiation Dynamics, Inc., Melville, NY (1976).
50. Clelland, M. R., *Lecture Notes on Electron Beam Processing, IAEA Regional Training Course*, Shanghai (1988).
51. Timmerman, U.S. Patent 3,142,629 to Radiation Dynamics, Inc. (July 1964).
52. ExxonMobil Chemical, *Vulcanization of Vistalon Polymers* (2000).
53. Tada, S., and Uda, I., *Radiat. Phys. Chem.*, 22, p. 575 (1983).
54. Sawasaki, T., and Nojiri, A., *Radiat. Phys. Chem.*, 31, p. 877 (1988).

55. Yongxiang, F., and Zueteh, M., in *Radiation Processing of Polymers* (Singh, A., and Silverman, J., Eds.), Hanser, Munich, p. 88 (1992).

56. Hunt, E. J., and Alliger, G., *Radiat. Phys. Chem.*, 14, p. 39 (1979).

57. Thorburn, B., and Hoshi, Y., *Meeting of Rubber Division of American Chemical Society*, Detroit, October 8–11, Paper 89 (1991).

58. Thorburn, B., *Meeting of Rubber Division of American Chemical Society*, Chicago, April 19–22, Paper 20 (1994).

59. Thorburn, B., and Hoshi, Y., Meeting of Rubber Division of American Chemical Society, Anaheim, CA, May 6–9, Paper 89 (1997).

60. Charlesby, A., *Nucleonics*, 12, p. 18 (1954).

61. Ivett, R. N., U.S. Patent 2,826,540 to Hercules Powder Co. (1958).

62. Precopio, E. M., and Gilbert, R., U.S. Patent 2,888,424 (1959).

63. Lewis, J. R., Japanese Patent Publication 40-25 351 to Hercules Powder Co. (1965).

64. Palmer, D. A., U.S. Patent 3,341,481 to Hercules, Inc. (1967).

65. Palmer, D. A., U.S. Patent Re 26 850 to Hercules, Inc. (1970).

66. Scott, H. G., U.S. Patent 3,646,155 to Midland Silicones Ltd. (1972).

67. *Di-Cup® and Vul-Cup® Peroxides Technical Data*, Bulletin PRC-102A, Hercules, Inc., Wilmington, DE (1979).

68. Griffin, J. D., Rubens, L. C., and Boyer, R. F., Japanese Patent Publication 35-13 138 (1960); British Patent 844-231 to Dow Chemical Co. (1960).

69. Okada, A., et al., U.S. Patent 3,542,702 to Toray Industries, Inc. (1970).

70. Atchinson, G., and Sundquist, D.J., U.S. Patent 3,852,177 to Dow Chemical Co. (1974).

71. Nojiri, A., Sawasaki, Y., and Koreda, T., U.S. Patent 4,367,185 to Furukawa Electric Co. Ltd. (1983).

72. Lohmar, E., and Wenneis, W., U.S. Patent 4,442,233 to Firma Carl Freudenberg (1984).

73. Nojiri, A., and Sawasaki, T., *Radiat. Phys. Chem.*, 26, p. 339 (1985).

74. ASTM D2765, Method A, Vol. 08.02, American Society for Testing Materials, Philadelphia, PA.

75. Benning, C. J., *J. Cell. Plast.* 3, p. 62 (1967).

76. Trageser, D. A., *Radiat. Phys. Chem.*, 9, p. 261 (1977).

77. Harayama, H., and Chiba, N., *Kino Zairo*, 2, p. 30 (1982).

78. Shina, N., Tsuchiya, M., and Nakae, H., *Japanese Plastic Age*, December, p. 37 (1972).

79. Sakamoto, I., and Mizusawa, K., *Radiat. Phys. Chem.*, 22, p. 947 (1983).

80. Shinohara, Y., Takahashi, T., and Yamaguchi, K., U.S. Patent 3,562,367 to Toray Industries, Inc. (1971).

81. Sagane, N., et al., U.S. Patent 3,711,584 to Sekisui Chemical Co. Ltd. (1973).

82. Park, C. P., in *Polymeric Foams* (Klempner, D., and Frisch, K. C., Eds.), Hanser, Munich, Chapter 9, p. 211 (1991).

83. Kiyono, H., et al., U.S. Patent 4,218,924 to Sekisui Chemical Co. (1980).

84. Park, C. P., in *Polymeric Foams* (Klempner, D., and Frisch, K. C., Eds.), Hanser, Munich, Chapter 9, p. 231 (1991).

85. Shinohara, Y., et al., U.S. Patent 3,562,367 to Toray Industries, Inc. (1971).

86. Tamai, I., and Yamaguchi, K., *Japanese Plastic Age*, December 22 (1972).

87. Park, C. P., in *Polymeric Foams* (Klempner, D., and Frisch, K. C., Eds.), Hanser, Munich, Chapter 9, p. 234 (1991).

88. Bradley, R., *Radiation Technology Handbook*, Marcel Dekker, New York, p. 247 (1983).

89. Cook, P. M., and Muchmore, R. W., U.S. Patent 3,086,242 to Raychem Corporation (1963).

90. Mehnert, R., Bogl, K.W., Helle, N., and Schreiber, G. A., in *Ullmann's Encyclopedia of Industrial Chemistry*, Vol. A22 (Elvers, B., Hawkins, S., Russey, W., and Schulz, G., Eds.), VCH Verlagsgesselschaft, Weinheim, p. 487 (1993).

91. Timmerman, R., U.S. Patent 3,296,344 to Radiation Dynamics, Inc. (1967).

92. Silverman, J., in *Radiation Processing of Polymers* (Singh, A., and Silverman, J., Eds.), Carl Hanser Verlag, Munich, Chapter 2, p. 20 (1992).

93. Bradley, R., *Radiation Technology Handbook*, Marcel Dekker, New York, p. 248 (1983).

94. Ellis, R. H., U.S. Patent 3,455,336 to Raychem Corporation (1969).

95. Derbyshire, R. L., U.S. Patent 4,287,011 to Radiation Dynamics, Inc. (1981).

96. Muchmore, R. W., U.S. Patent 3,542,077 to Raychem Corporation (1970).

97. Dyer, D. P., Tysinger, A. D., and Elliott, J. E., U.S. Patent 6,326,550 to General Dynamics Advanced Technology Systems, Inc. (2001).

98. Brax, H. J., Porinchak, J. F., and Weinberg, A. S., U.S. Patent 3,741,253 to W. R. Grace & Co. (1973).

99. Mueller, W. B., et al., U.S. Patent 4,188,443 to W. R. Grace & Co. (1980).

100. Bornstein, N. D., et al., U.S. Patent 4,064,296 to W. R. Grace & Co. (1977).

101. Kupczyk, A., and Heinze, V., U.S. Patent 5,250,332 to RXS Schrumpftechnik Garnituren G.m.b.H, Germany (1993).

102. Bax, S., Ciocca, P., and Mumpower, E. L., U.S. Patent 6,150,011 to Cryovac, Inc. (2000).

103. Koleske, J. V., *Radiation Curing of Coatings*, ASTM International, West Conshohocken, PA, p. 213 (2002).

104. Santorusso, T. M., *Radiation Curing*, 11, No. 3, August, p. 4 (1983); *Proceedings, Radcure '84*, Atlanta, GA, September 10–13, p. 16 (1984).

105. Zillioux, R. M., *Proceedings of Radcure '86*, Baltimore, MD, September 8–11, p. 8 (1986).

106. Mehnert, R., Pinkus, A., Janorsky, I., Stowe, R., and Berejka, A., *UV&EB Curing Technology and Equipment*, Vol. 1, John Wiley & Sons Ltd., Chichester/SITA Technology Ltd., London, p. 34 (1978).

107. Maguire, E. F., *RadTech Report*, 12, No. 5, September/October, p. 18 (1998).

108. Seidel, J. R., in *Radiation Curing of Polymers* (Randell, D. R., Ed.), The Royal Society of Chemistry, London, p. 12 (1987).

109. Mehnert, R., Bogl, K. W., Helle, N., and Schreiber, G. A., in *Ullmann's Encyclopedia of Industrial Chemistry*, Vol. A22 (Elvers, B., Hawkins, S., Russey, W., and Schulz, G., Eds.), VCH Verlagsgesselschaft, Weinheim, p. 485 (1993).

110. Hara, K. J., in *Radiation Curing of Polymers* (Randell, D. R., Ed.), The Royal Society of Chemistry, London, p. 127 (1987).

111. Biro, D. A., *RadTech Report*, 16, No. 2, March/April, p. 22 (2002).

112. Gamble, A. A., in *Radiation Curing of Polymers* (Randell, D. R., Ed.), The Royal Society of Chemistry, London, p. 76 (1987).

113. Lapin, S. C., *RadTech Report*, 22, No. 5, September/October, p. 27 (2008).

114. Meij, R., *RadTech Europe 2005, Conference Proceedings*, Barcelona, October (2005).

115. Sanders, R., *RadTech Report*, 17, No. 5, September/October, p. 30 (2003).

116. Chrusciel, J., *Proceedings RadTech Europe 2001*, Basel, Switzerland, October 8–10, p. 373 (2001).
117. Fisher, R., *Conference Proceedings, TAPPI Hot Melt Symposium 1999*, Durango, CO, June 13–16, p. 115 (1999).
118. Miller, H. C., *J. Adhesive Sealant Council*, p. 111 (1998).
119. Ramharack, R., et al., *Proceedings RadTech North America '96*, April 28–May 2, p. 493 (1996).
120. Nitzl, K., *European Adhesives & Sealants*, 13, No. 4, December, p. 7 (1996).
121. Ramharack, R., et al., *Adhesives Age*, 39, No. 13, December, p. 40 (1996).
122. Rangwalla, I., and Maguire, E. F., *RadTech Report*, 14, No. 3, May/June, p. 27 (2000).
123. Lapin, S. C., *RadTech Report*, 15, No. 4, July/August, p. 32 (2001).
124. Henke, G., *Proceedings RadTech Europe 2001*, Basel, Switzerland, October 8–10, p. 361 (2001).
125. Chaix, C., *RadTech Report*, 11, No. 1, January/February, p. 12 (1997).
126. Walton, T. C., and Crivello, J. V., *Conference Proceedings, '95 International Conference on Composite Materials and Energy*, Montreal, Canada, May 8–10, p. 395 (1995).
127. Singh, A., et al., *Conference Proceedings, '95 International Conference on Composite Materials and Energy*, Montreal, Canada, May 8–10, p. 389 (1995).
128. Walton, T. C., and Crivello, J. V., *Materials Challenge—Diversification and the Future*, Vol. 40, Book 2, Anaheim, CA, May 8–11, p. 1266 (1995).
129. Guasti, F., Matticari, G., and Rossi, E., *SAMPE J.*, 34, p. 29 (1998).
130. Guasti, F., and Rossi, E., *Composites A Appl. Sci. Manuf.*, 28A, No. 11, p. 965 (1997).
131. Raghavan, J., and Baillie, M. R., *Conference Proceedings, Polymer Composites '99*, Quebec, Canada, October 6–8, p. 351 (1999).
132. Crivello, J. V., Walton, T. C., and Malik, R., *Chemistry of Materials*, 9, No. 5, May, p. 1273 (1997).
133. Hill, S., *Materials World*, 7, July, p. 398 (1999).
134. Beziers, D., et al., U.S. Patent 5,585,417 to Aerospatiale Societe Nationale Industrielle (1996).
135. Berejka, A. J., *RadTech Report*, 16, No. 2, March/April, p. 33 (2002).
136. Lopata, V. J., and Sidwell, D. R., *RadTech Repair*, 17, No. 5, September/October, p. 32 (2003).
137. Bhattacharyya, D., Pillai, K., Oliver, M. R., Chyan, T. L., and Timmons, R. B., *Chem Mater.*, 19, p. 2222 (2007).
138. L'Anunziata, M. F., *Radioactivity: Introduction and History*, Elsevier, Amsterdam (2007).
139. Rosiak, J. M., Janik, I., Kadlubowski, S., Kozicki, M., Kujawa, P., Stasica, P., and Ulanski, P., *Nucl. Instrum. Meth. Phys. Res. B*, 208, p. 325 (2003).
140. El Salmawi, K. M., *J. Macromol. Sci. A*, 44, June, p. 619 (2007).
141. Davidson, R. S., *Radiation Curing*, Report 136, 12, No. 4, November, Rapra Technology Ltd., p. 27 (2001).
142. Ebnesajjad, S. and Morgan R.A., *Fluoropolymer Additives*, Elsevier, Oxford, UK, p. 39 (2012).
143. Drobny, J.G., *Ionizing Radiation and Polymers*, Elsevier, Oxford, UK, p. 196 (2013).
144. Drobny, J.G., *Ionizing Radiation and Polymers*, Elsevier, Oxford, UK, p. 204 (2013).

10

Dosimetry and Radiometry

The measurement of the amount of radiant energy delivered by the source is important in order to determine if the equipment operates properly and to ensure that the product will be of required quality. Because of the fundamental difference between UV and EB sources, the measuring methods and instruments are basically different.

10.1 EB Dosimetry

Industrial applications of ionizing radiation encompass a wide range of absorbed doses (about six orders of magnitude), dose rates (about twelve orders of magnitude), and energies (over two orders of magnitude).[1]

The absorbed dose is the quantity of interest to be measured in order to evaluate the effects of the radiation on the processed materials. The methods used for that are known as *dosimetry*. There are a multitude of established dosimetric methods, and some of them are listed in Table 10.1.

The main functions of dosimetry are[1]:

- The dose measurement leading to successful processing with a sufficient uniformity
- Process validation, including dose settings for meeting process specifications
- Process qualification, including the establishment and locations and values of minimum and maximum doses
- Process verification, by the coordination of routine dosimetry and the monitoring of process parameters
- Proper calibration and traceability to standards using reference dosimetry

The choice of the method depends on the type of process. With electron beams, the critical process parameters are beam energy, beam current, scanning factors and uniformity, beam pulse characteristics, and the configuration of the product being processed.

TABLE 10.1

Examples of Dosimetry Systems

Dosimeter Type	Type of Readout	Examples	Absorbed Dose Range, Gy
Calorimeters	Temperature measurements	Graphite, water, polystyrene	10^1–10^4
Radiochromic films	Spectrophotometers	Dyed and clear plastic films	10^0–10^6
Inorganic crystals	EPR[a] spectra	LiF, SiO_2	10^3–10^7
Organic crystals	EPR[a] spectra	Alanine, sucrose, cellulose	10^0–10^5
Chemical solutions	Spectrophotometers	Ceric-cerous, organic acids	10^2–10^5
Semiconductors	Electric measurements	Si diodes	10^0–10^4

[a] EPR—Electron paramagnetic resonance.

The most convenient and widely used method is *film dosimetry*, which has been used for several decades. Commercially available film dosimeters are essentially thin plastic films either clear or containing dyes or dye precursors. These so-called *radiochromic films* are available in thicknesses ranging from 0.005 to 1 mm and are used to monitor electron beams and gamma ray doses from 10 to 10^5 Gy.[2] These doses are typical for medical applications; radiation curing of coatings, adhesives, wire, and cable insulations; cross-linking of shrinkable tubing and films and composites; etc. When irradiated, radiochromic films change irreversibly their optical absorbance in proportion to the absorbed dose (see also Section 10.2.3). To minimize errors due to variation in the film thickness, the dosimeter response is usually expressed as the radiation-induced change in absorbance divided by the dosimeter film thickness (see Figure 10.1). A partial list of currently available EB dosimeters is shown in Table 10.2. The appropriate test procedures are described in several ASTM standards.[3]

There are several dosimeter systems being used for determining the absorbed dose. The simplest ones are *thin-film dosimeters*, which are suitable for use in routine practice.

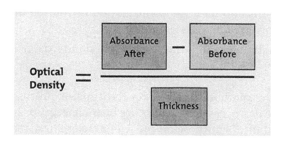

FIGURE 10.1

Equation for the change in optical density of film dosimeters.

TABLE 10.2

Examples of Available Thin-Film EB Dosimeters

Type	Film Thickness, μm	Analysis Wavelength, nm	Dose Range, kGy
Radiachromic films with special dyes			
in polyamide (nylon)	50 or 10	600–510	0.5–200
in polychlorostyrene	50	630 or 430	1–300
in polyvinyl butyral	22	533	1–200
Radiachromic microcrystalline layer on a 100 μm polyester support	6 (sensor)	650 or 400	0.1–50
Cellulose triacetate	38 or 125	280	5–300
Dyed cellulose diacetate	130	390–450	10–500
Dyed (blue) cellophane	20–30	650	5–300

In general, readings of any dosimeter can be influenced by the following factors[4]:

- Dose rate or dose fractionation
- Temperature, relative humidity, oxygen content, light
- Its stability after irradiation

Each type of dosimeter requires a specific procedure to ensure accurate and reproducible results, such as postirradiation heat treatment. Others need to be stabilized for a certain time (up to 24 h with some) before readings of absorbance are taken.[5-7] The absorbance reading can be done by conventional spectrophotometry or other, more involved methods.

The original radiochromic films were blue cellophane films colored by a blue diazo dye that could be bleached by ionizing radiation, with thicknesses between 19 and 26 μm, depending on the manufacturer. Such thin transparent cellophane films have been used as decorative heat-sealable and moisture-resistant packaging films. This might have been one of the reasons why this blue film was found not to be sufficiently reproducible in response to the radiation.[8] Improved cellophane films are produced specifically for dosimetry in two different intense colors, orange and violet, with dissolved disazo dyes. They are delivered in thickness of 1 mil (25.6 μm) and maximum absorptions of 440 and 560 nm, respectively. Bleaching occurs gradually over the dose range of 10–300 kGy.[2]

Other commercially radiochromic films are poly(methyl methacrylate) (PMMA), cellulose triacetate (CTA),[1,8] polyamide (nylon) films, poly(vinyl butyrate) with pararosaniline and p-nitrobenzoic acid,[1] and alanine films.[2] The CTA films are undyed; PMMA films are available undyed and dyed (red and amber).[2]

Thin-film dosimeters are commonly used for the following measurements[9]:

- The *surface area rate* (or *processing coefficient*), relating the area irradiated per unit time to the beam current, and the absorbed dose are determined by measuring the surface dose at several beam current levels.
- The *cure yield* is calculated in terms of the average absorbed dose delivered to the layer of interest multiplied by the conveyor speed per unit effective beam current. It can be estimated by measuring the dose over the range of beam currents at different speeds.
- The *dose uniformity across the width of the beam* is measured by placing the film chips at about 1 in. (25 mm) intervals or using a long strip of film across the entire width of the beam.
- The *depth-dose distribution* is measured by irradiating a stack of radiachromic film chips with a thickness slightly greater than the practical range at the energy of interest. The depth dose is determined by evaluating the individual chips.

A dose reader designed for direct reading from exposed films, offered by Elektron Crosslinking AB of Sweden, displays dose values directly in kGy (see Figure 10.2). The reader can be used for continuous measurement with the connection through the serial port to a PC. The complete set for continuous measurements is shown in Figure 10.3. The features of this dose reader are listed in Table 10.3. A recently developed dose reader from the same company using a long film strip provides in addition the dose distribution over the entire width of the processed product. The type of the film used in DR-030 is the same as that used in DR-020. Dose reader DR-030 is shown in Figure 10.4. The company will supply the new reader in a similar set as the DR-020 shown in Figure 10.3.

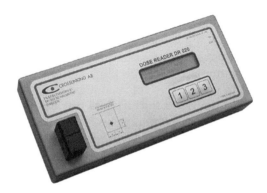

FIGURE 10.2
Dose reader DR-020 for irradiated films. (Courtesy of Elektron Crosslinking AB.)

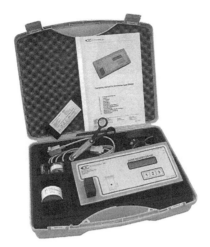

FIGURE 10.3
Complete dose reader set for continuous measurement. (Courtesy of Elektron Crosslinking AB.)

TABLE 10.3

Features of Dose Reader DR020

Weight	1.9 kg (4.19 lb)
Dimensions (L × W × H)	295 × 135 × 57 mm (11.6 × 5.3 × 2.2 in.)
Power supply	8–12 V DC
Keyboard	3 mechanical keys covered with keyboard overlay
Display	Dot matrix, 20 characters, 2 lines
Interface	Serial interface, data output via null modem cable
Electronics	Card-size processor with 10 bits A/D converter
Dose range	5 kGy–60 (75) kGy
Thickness of measuring film	20, 10, 5 μm
Applicable standard	Dose mapping and routine dosimetry according to ISO/ASTM 51275

To ensure accuracy and reliable data, certain precautions must be taken when using radiochromic films.[2]

- They should not be irradiated in extremes of relative humidity, i.e., below 20% and above 80%. If such conditions cannot be avoided during irradiation, then the films should be sealed in polyethylene pouches with the relative humidity between 50% and 70% before irradiation.
- Since the films are read optically, they should be kept free of dirt, scratches, and fingerprints.

FIGURE 10.4
Newer model of dose reader for irradiated films DR-030. (Courtesy of Elektron Crosslinking AB.)

- The films must be protected from UV irradiation at all times.
- In general, the films should not be read until approximately 24 h after irradiation because the full dye development requires several hours.

Any dosimeter used to determine absorbed dose in an irradiated product has to be calibrated.[1,2] The adiabatic character of electron beam deposition is used in calorimetry, which is the primary absolute method of measuring the absorbed dose (energy per unit mass).[4] An example of the instrument for this purpose is the graphite calorimeter developed in Risø National Laboratory in Denmark.[10,11] This calorimeter (see Figure 10.5) is reported to be suitable for electrons from a linear accelerator with energies higher than 5 MeV and shows an accuracy of ±2%.[12] Other types of instruments for this purpose are the water

FIGURE 10.5
Graphite calorimeter. (Courtesy of Risø National Laboratory.)

calorimeter[13] and polystyrene calorimeter.[14] For example, a totally absorbing graphite body calorimeter is reported to be suitable for measuring energy-deposited electron accelerators operating within the range of 4–400 MeV.[15–17]

Essentially, a calorimeter measures absorbed dose in terms of the radiation-induced temperature rise in the calorimeter core connected electrically to the outside. The dose to the dosimeter being calibrated is related to the associated calorimeter reading of the net temperature rise. These calorimeters are used as reference dosimeters for high-energy electron beams (2–12 MeV) and also for small references and routine dosimeters (pellets and radiochromic films).

The calibration is done by national or secondary standards laboratories. Calibration at the National Institute of Standards and Technology (NIST) involves exposure to either gamma ray source (^{60}Co) over the dose range of 10^0–10^7 Gy, or high-energy electron beam (1–28 MeV) combined with a graphite or water calorimeter.[1]

Alanine dosimeters are based on the ability of l-α alanine (a crystalline amino acid) to form a very stable free radical when subjected to ionizing radiation. The alanine free radical yields an electron paramagnetic resonance (EPR) signal that is dose dependent yet independent of the dose rate, energy type, and relatively insensitive to temperature and humidity.[18] Alanine dosimeters are available in the form of pellets or films and can be used for doses ranging from 10 Gy to 200 kGy.[19] A reference calibration service using the alanine EPR system was developed, and the scans were sent to the service center by mail. Currently, the available system allows transferring the EPR scan to a NIST server for a calibration certificate. This way the procedure has been shortened from days to hours.[19]

Real-time monitors provide a continuous display in real time of dose delivered to product, as well as the energy of the electron beam. Moreover, they can show the variations in dose across a wide web if multiple detectors are installed and provide alarm signals to warn the operator of high- and low-dose conditions, and they record the performance of the processor for production control, quality assurance, and maintenance needs. The major advantage of these real-time instruments is that they monitor the current and beam energy independently.[20,21]

In real industrial settings, electron beam processors use dosimeters to evaluate yield measurement for a specific machine, beam uniformity (distribution across the width), and depth dose.[22]

10.2 UV Radiometry

There are several devices measuring either the number of photons (total or per unit time) or the amount of radiation converted into electric signal or into heat. These are *actinometers* and *radiometers*.

10.2.1 Actinometers

Actinometers are chemical systems or physical devices that determine the number of photons in a beam either integrally or per unit time. In a *chemical* actinometer, the photochemical change can be directly related to the number of photons absorbed, while the *physical* device gives a reading correlated to the number of photons detected.[23]

10.2.2 Radiometers

Radiometers are quantitative physical detectors of radiation, which convert the radiant energy to an electrical signal. They can be used for monitoring and controlling UV curing systems in order to produce the desired cure.

In their simplest form, radiometers monitor irradiance (in W/cm^2) and radiant energy density (in J/cm^2) for the bandwidth of the instrument. In addition, profiling radiometers can provide irradiance profiles as a function of time. The results from the monitoring of a process can be effectively used to correlate exposure conditions to the physical properties of the cured product. If needed, they can also become the specifications of exposure in the design of production systems. Usually, radiometers are placed in the same position as the material that is being cured.

10.2.2.1 Radiometers for Conventional Ultraviolet Lamps

Some self-contained instruments are designed for online measurements in UV curing units to determine the integral UV dose delivered to the material being cured, peak irradiance, and irradiance/temperature profiles.[4] The quantities and units of measurements have been established based on IUPAC's recommendation of 1996.[23] These are listed in Table 10.4.

Depending on their function, radiometers can be either thermal or photon detectors. In *thermal detectors*, the incident photon energy is converted into heat, which is then measured. The measured data are independent of wavelength. *Photon detectors* are based on photoelectric effect and measure spectrum intensity. The results are dependent on wavelength.

Instrument manufacturers offer a variety of devices, including monitoring systems for standard process or for spot curing technology and systems for process control. More advanced systems "map" the entire process and evaluate the lamp focus, reflector focus, and reflector efficiency within a specific region, and compare spectral output, degradation, and uniformity of one or more lamps in multiple-lamp arrays. Such a system also measures and documents substrate temperatures during the entire process and tracks and stores data for statistical quality control, regulation compliance, or historical comparisons[24]; detects changes; and sounds the alarm when the UV drops below a set limit.[25] Examples of commonly used radiometers are listed in Table 10.5.

TABLE 10.4

Quantities and Units Related to UV Radiometry

Quantity	Symbol	SI Unit	Common Unit
Irradiance	E	W m^{-2}	—
Photon exposure	H$_p$	m^{-2} mol m^{-2}	—
Photon flow	Φ_p	s^{-1} mol s^{-1}*	—
Photon irradiance Photon flux	E$_p$	m^{-2} s^{-1} mol m^{-2}*	—
Radiant energy	Q	J	—
Radiant power	P	W	—
Radiant exposure	H	J m^{-2}	
Spectral irradiance	E$_\lambda$	W m^{-3}	W m^{-2} nm^{-1}
Spectral photon flow	$\Phi_{p\lambda}$	s^{-1} m^{-1} mol^{-1} s^{-1} m^{-1}*	s^{-1} nm^{-1} mol^{-1} s^{-1} nm^{-1}*
Spectral photon flux Photon irradiance	E$_{p\lambda}$	s^{-1} m^{-3} mol s^{-1} m^{-3}*	s^{-1} m^{-2} nm^{-1} mol s^{-1} m^{-2} nm^{-1}*
Spectral radiant power	P$_\lambda$	W m^{-1}	W nm^{-1}

* If the amount of photons is used

TABLE 10.5

Examples of Radiometers for UV Measurement

Instrument	Spectral Range, nm	Measuring Range[a]
UV intensity meters	250–260 280–320 320–390	0–19 W/cm^2
Spot cure radiometers	250–445	0–19.99 W/cm^2
Radiometers for xenon flash	326–401	57 nW/cm^2–57 W/cm^2
UV cure radiometers	250–400	1 nW/cm^2–1 W/cm^2 150 nW/cm^2–0.5 W/cm^2
Handheld radiometers	250–290 280–340 335–380 (Three sensors) 240–425 (in 1 nm steps)	0–20 m W/cm^2 0–200 W/cm^2 0–2000 µW/cm^2 0–200 mW/cm^2 Min. 10 nW/cm^2/nm Max. 10 W/cm^2

[a] Irradiance or spectral irradiance

Radiometers are exposed to the UV source and provide the total energy (J/cm^2) or UV peak irradiance (W/cm^2) in absolute units. Depending on the design, they provide data in one or four spectral bandwidths. They are used in all UV curing chemistries, including inks, adhesives, solder masks, and epoxies. Besides measuring the performance of a UV lamp, they compare the

FIGURE 10.6
Example of an absolute reading radiometer displaying four different bandwidths simultaneously. (Courtesy of EIT Inc.)

efficiency of curing systems and establish the UV level for proper curing. An
example of a general-purpose radiometer is shown in Figure 10.6.

Radiometers for spot cure monitor the performance (irradiance), measure the
light guide degradation, determine optimum positioning of the light guide
cable, and compare the efficiency of spot curing systems. An example of a
spot cure radiometer is shown in Figure 10.7. A radiometer measuring both
spot and flood cures is shown in Figure 10.8.

FIGURE 10.7
Monitor for spot curing. (Courtesy of EIT Inc.)

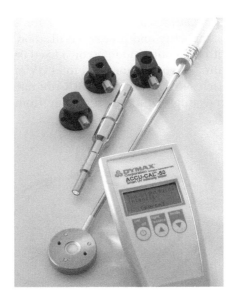

FIGURE 10.8
Radiometer for spot and flood cures. (Courtesy of Dymax Corporation.)

Profiling radiometers measure and display the peak power and total density ion of a UV curing system and also profile the temperature and irradiance as a function of time. The information is transferred into a computer. These radiometers are capable of comparing characteristics of multiple lamps, comparing UV systems over time, or comparing different systems to each other. They track and store archival data. An example of profiling radiometers is shown in Figure 10.9.

FIGURE 10.9
Profiling radiometers. (Courtesy of EIT Inc.)

Radiometers for three-dimensional cure are used for simultaneous multi-point measurements, for setup and process verification of the lamp system. They can be used with UV lamps mounted on a fixed bank or a robotic arm. The collected exposure data (irradiance and total UV energy) are displayed on a computer for each sensor position. A 3DCure radiometer is shown in Figure 10.10, and an example of a screen display from a 3D radiometer is shown in Figure 10.11.

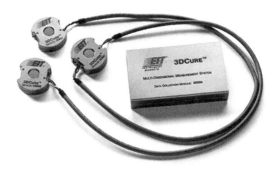

FIGURE 10.10
Radiometer for 3DCure. (Courtesy of EIT Inc.)

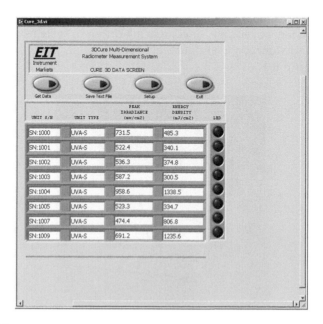

FIGURE 10.11
Display from 3DCure radiometer. (Courtesy of EIT Inc.)

Online monitoring equipment provides continuous monitoring of UV lamp intensity. Results are displayed as a percentage of the original output, which is usually set at 100%. Equipment with compact sensors for monitoring is shown in Figure 10.12.

Probes (example shown in Figure 10.13) are used for applications where space is limited or difficult to access. They are capable of establishing and maintaining a process window and coordinate readings from online displays.

Equipment manufacturers suggest sensor locations as shown in Figure 10.14. However, since lamps vary in design and installation, these locations may in some cases be different.

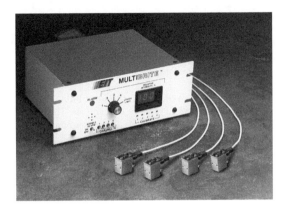

FIGURE 10.12
Online monitoring instrumentation. (Courtesy EIT Inc.)

FIGURE 10.13
Probe for production ambient light measurement. (Courtesy of EIT Inc.)

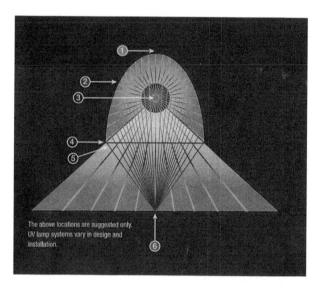

The above locations are suggested only.
UV lamp systems vary in design and
installation.

FIGURE 10.14

Suggested sensor location. 1—Behind reflector near end of bulb, 2—Behind reflector looking at bulb and reflected energy, 3—From end of lamp housing, 4—Through filter material or rod, 5—Looking up at bulb and reflector from below, 6—Directly under the lamp at substrate level. (Courtesy of EIT Inc.)

10.2.2.2 Radiometers for Ultraviolet Light-Emitting Diode Lamps

Conventional medium-pressure broadband UV lamps emit radiation across a broad electromagnetic spectrum. Output from these types of sources includes UV, visible, and infrared (IR) energy.

UV-LEDs are narrowband sources. Production UV-LED sources have their spectral emissions somewhere in the 365–405 plus nm region. UV-LEDs are described and identified by their most dominant (e.g., 395 nm) spectral output. If measured with a spectral radiometer, the user would see that the manufacturer has binned the individual LED chips so that the most intense UV output is clustered around the dominant name line (i.e., 395 nm if the source is a 395 nm source).[26] Comparison of power output versus wavelength for arc mercury lamps and UV-LED lamps is shown in Figure 10.15.[27]

Measuring and characterizing LEDs requires an understanding of the UV source and the UV measurement instrument. Instruments used to measure broadband, medium-pressure UV microwave, or arc lamps may or may not be suitable for use with UV-LED sources.

On products, manufactured with UV-LED, the LED is extremely close to the product, and the process is optimized for a high-intensity, narrowband width (monochromatic) UV source. In contrast to products manufactured with a broadband (arc) lamp, the source is set up for far-field distant curing with low intensity levels.[28]

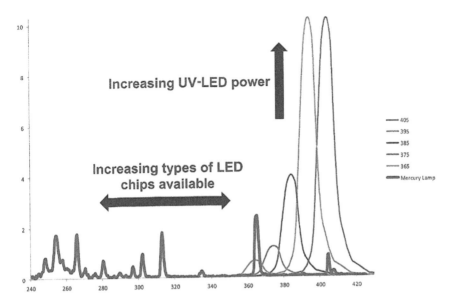

FIGURE 10.15
Comparison of power output and wavelength for arc mercury light and UV-LED lamps.
(Courtesy of EIT Inc.)

UV-LEDs sold for industrial curing applications, typically specify a center wavelength (CWL) of ±5 nm. Therefore, the CWL for a 395 nm source will fall between 390 and 400 nm. The actual distribution of UV energy at all power levels is determined by how the diodes are selected or "binned." Based on the energy distribution, instrument responsivity of a radiometer for a 395 nm LED would need to be 370–422 nm with a flat response across the top of the band.

The instrument used for the measurement, a *profiling radiometer* (see Figure 10.16), displays information in the same way as radiometers used for broadband sources. The LED irradiance (W/cm²), energy density (J/cm²), and irradiance profile (Watts/cm² as a function of time) are available on the display of the instrument. LEDCure radiometers (EIT, Inc.) feature a patented "Total Measured Optical Response" that includes all optical components in the overall response of the instrument, not just the optical filter. The instrument response, which has been optimized for UV-LEDs, is designated with an "L" (LED) band response followed by the center wavelength of the LED.

The EIT PowerMAP II is a profiling radiometer that provides the irradiance (W/cm²), energy density (J/cm²), irradiance profile (Watts/cm² as a function of time), and temperature profile (°C as a function of time); it is a new instrument from EIT. This compact, one-piece instrument is 60% smaller than the original EIT PowerMAP and measures UV in four (UVA, UVB, UVC, UVV) EIT spectral regions. The product features larger internal

FIGURE 10.16
Profiling radiometer for UV-LED lamps. (Courtesy of EIT Inc.)

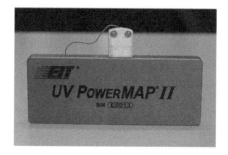

FIGURE 10.17
PowerMAP® profiling radiometer. (Courtesy of EIT Inc.)

memory for more data gathering, and rapid transfer of data to the new UV PowerView Software III graphic data analysis program via USB interface. The PowerMAPII radiometer is shown in Figure 10.17.

Applications with fast process speeds such as digital LED printers require a radiometer with a fast, effective sample rate to accurately measure the irradiance (Watts/cm²), energy density (Joules/cm²) and irradiance profile (Watts/cm² as a function of time). The EIT LEDCure® family of UV measurement solutions continues to grow in response to the needs of source suppliers, equipment integrators, formulators, and end users using UV-LEDs. Products described in this update are new and/or are expected to be commercially available very soon. EIT® LEDMAP™ utilizes EIT's POWERMAP®/POWERMAP® II technology, which has over 22 years of fast (2048 Hz) sample rate experience. The LEDMAP™ allows you to measure and "see" individual LED passes at speeds up to 400 ft (122 m) per minute. The first commercially available LEDMAP™ will have the L-395 band with EIT's patented Total Measured Optic Response™ for accurate and repeatable measurements. The LEDMAP™ can collect up to 65 min of data and has an option to also measure and profile process temperature. The collected data is transferred via USB to EIT's UV PowerView® III Software. PowerView III allows for analysis, troubleshooting, and comparison of the same or different LED systems. The LEDMAP™ sample rate provides enough resolution to see individual LED passes at fast speeds. The data can also be exported to Excel.[29]

A continuous instrumentation for monitoring UV lamps is shown in Figure 10.18.

10.2.3 Radiochromic Films

Radiochromic films are used routinely in the measurement of EB doses in the form of cellophane strips or tabs that attach to the surface of the irradiated web or part (see Section 10.1) but can also be used in the UV curing process.[30] They respond to total time-integrated energy by changing color or optical density, typically in the wavelength range from 200 to 350 nm, without any wavelength specificity. Their advantage is that they can pass through the UV system, including nips of rollers, without being damaged. Their drawback is that they respond and record accumulated energy only and cannot distinguish between exposures of individual lamps if used in multilamp systems.

Obviously, the user has a large selection of instruments, each of them having a different function. The selection depends on what the user wants to measure and for what purpose. In *process design*, the main objective[30] is to

FIGURE 10.18
Online monitoring of UV lamps. (Courtesy of ESEN UV&EB Technology.)

optimize the cured properties of the finished product. To that end, the following conditions have to be established and quantified:

- Effective irradiance or profile
- Spectral distribution
- Time or speed (to determine energy)
- Amount of infrared energy or temperature

Thus, these measurements provide quantitative information about the critical parameters used in the process and help establish the limits within which the process is successful, or the so-called process window.

Once the process window is established, the goal is to maintain the operating condition within the established limits. This is often referred to as *process control*. The primary purpose of process control is to monitor the process to get a feedback by means of online radiometric measurements and take action to keep it within the established limits. Process monitoring has to verify that the key process variables remain within the specified limits and to interpret changes in the exposure conditions to help maintain control. Once established, proper measurements are invaluable in monitoring the condition of the UV lamps and determining when they have to be maintained or replaced.[4]

Certain processes, such as 3D curing, roll-to-roll printing and coating, and jet printing, are difficult or even impossible to evaluate by the use of traditional radiometry. On the other hand, radiochromic films exhibit a change of hue with exposure, changing their optical density in a specific color component, but cannot resolve irradiance or any information on the irradiance profile of exposure. An advanced study to solve some of the problems by evaluating methods of correlation to produce a numerical measure of UV exposure was performed. The principal purpose of this study was to explore the use of standard instruments to quantify the response of radiochromic films in terms of transmission or reflection densitometry. It concluded that several radiochromic films have a wide dynamic exposure range, good linearity, and nearly perfect cosine response, are economical, and can be read with comparatively inexpensive instruments. This allows correlation of their optical density with instrument radiometry for process design optimization or periodic measurements to verify lamp condition over time. The important conclusion was that when carefully designed and used, radiochromic films can be a useful extension of—but not a substitute for—instrument radiometry.[31]

The subject of EB and UV dosimetry is discussed thoroughly in Mehnert et al.[4] A comprehensive coverage of main instrumentations for UV radiation measurements is in *Radiometers*, Document 100200, CAT REV B 305 (EIT Instrument Market, Sterling, VA).[26,27]

References

1. McLaughlin, W. L., and Desrosiers, M. F., *Radiat. Phys. Chem.*, 46, p. 1163 (1995).
2. Humphreys, J. C., and McLaughlin, W. L., *IEEE Trans. Nucl. Sci.*, NS-28, p. 1797 (1981).
3. ASTM Standards E170, E668, E1026, E1204, E1261, E1275, E1276. Additional source: ASTM Standards E1204, E1261, E1275, and E1276 were replaced by Standards ISO/ASTM51204, ISO/ASTM51261, ISO/ASTM51275, and ISO/ASTM51276, respectively.
4. Mehnert, R., Pincus, A., Janorsky, I., Stowe, R., and Berejka, A., *UV and EB Technology and Equipment*, Vol. 1, John Wiley & Sons, Chichester/SITA Technologies Ltd., London, p. 107 (1998).
5. Janovsky, I., and Mehta, K., *Radiat. Phys. Chem.*, 43, p. 407 (1994).
6. Mc Laughlin, W. L., Puhl, J. M., and Miller, A., *Radiat. Phys. Chem.*, 46, p. 1227 (1995).
7. Abdel-Fattah, A. A., and Miller, A., *Radiat. Phys. Chem.*, 47, p. 611 (1996).
8. McLaughlin, W. L., *Radiat. Phys. Chem.*, 67, p. 561 (2003).
9. Rangwalla, I., and Swain, M., *RadTech Report*, 17, No. 5, September/October, p. 26 (2003).
10. Brynjolfsson, A., and Thaarup, G., *Rep. Riso*, 53, Riso National Laboratory, Denmark (1963).
11. Holm, N. W., and Berry, R. J., *Manual on Radiation Dosimetry*, Marcel Dekker, New York (1970).
12. Zagórsky, Z. P., in *Radiation Processing of Polymers* (Singh, A., and Silverman, J., Eds.), Carl Hanser Verlag, Munich, p. 279 (1992).
13. McLaughlin, W. L., Walker, M. L., and Humphreys, J. L., *Radiat. Phys. Chem.*, 46, No. 4–6, September, p. 1235 (1995).
14. Miller, A., Kovacs, A., and Kuntz, F., *Radiat. Phys. Chem.*, 46, No. 4–6, September, p. 1243 (1995).
15. McLaughlin, W. L., et al., *Dosimetry for Radiation Processing*, Taylor & Francis Group, London (1989).
16. Sunaga, H., et al., in *Proceedings of the 6th Japan-China Bilateral Symposium on Radiation Chemistry*, Waseda University (November, Tokyo, Japan) (1994).
17. Janovsky, I., and Miller, A., *Appl. Radiat. Isot.*, 38, p. 931 (1987).
18. Garcia, R. M. D., et al., *Radiat. Phys. Chem.*, 71, p. 373 (2004).
19. *e-scan™ Alanine Dosimetry System*, Brochure, Bruker Biospin, GmbH, www.bruker-biospin.com.
20. Kneeland, D. R., Nablo, S. V., Weiss, D. E., and Sinz, T. E., *Radiat. Phys. Chem.*, 55, p. 429 (1999).
21. Korenev, S., Korenev, I., Rumega, S., and Grossman, L., *Radiat. Phys. Chem.*, 71, p. 315 (2004).
22. Rangwalla, I., *UV+EB Technology*, Vol. 3, No. 2, p. 10 (2017).
23. Commission on Photochemistry, Organic Chemistry Division, International Union of Applied Chemistry, *Pure Appl. Chem.*, 68, p. 2223 (1996).
24. Braun, A. M., Maurette, M.-T., and Oliveros, E., *Photochemical Technology*, John Wiley & Sons, Chichester, p. 51 (1991).

25. *Guide to UV Measurement: Technology, Radiometry and Measurement for Industrial UV Curing*, RadTech International North America, Chevy Chase, MD, www.radtech.org.
26. Raymont, J. and Kashyap, A., *UV Lead Curing Technology*, February 2016, RadTech International, p. 48.
27. Raymont, J., May, J., Lawrence, M., UV/EB West 2017, in *Conference Proceedings* (February 27–March 1, San Francisco), RadTech International, www.radtech.org.
28. Raymont, J., *UV+EB Technology*, Vol. 2, No. 2, p. 8 (2019).
29. EIT LEDCURE™ UV Radiometers, Brochure P/N IM- 01104, Rev. B, August 2019, EIT, Inc., www.eit.com.
30. Diehl, D., and Abbe, P., *RadTech Report*, 14, No. 1, January–February, p. 14 (2000).
31. Stowe, R. W., in *Technical Conference Proceedings, RadTech 2002 North America*, (April 28–May 1, Indianapolis, IN), p. 475 (2002).

11

Safety and Hygiene

Ultraviolet radiation and electron beam technology has been recognized as a process successfully using low or zero content of volatile organic compounds (VOCs) for more than 30 years. This is a very important advantage over traditional processes in coating, printing, and adhesive applications using organic solvents and other volatile ingredients exclusively. However, as with any industrial manufacturing process, there are disadvantages and some hazards due to equipment, procedures, and raw materials used. In UV/EB curing, these hazards are attributable to reactive chemicals, some volatiles, and the nature of the energy sources.

Ultraviolet energy used in the curing process is a nonionizing or *actinic* form of electromagnetic radiation. The energy level is not high enough to penetrate into the human body and interact with the tissues causing cell damage, as do other forms of radiation. However, a direct exposure to UV light can have effects on both skin and eyes.

Electron beam processors generate two types of *ionizing* radiation: their primary product is high-energy electrons, and their secondary product is X-rays resulting from their interaction with matter. The ionizing radiation is damaging because of its capability of penetration into the human body.

The radiation curable liquid systems consist of a variety of chemical compounds, some of which may pose hazards due to their toxicity or tendency to irritate skin or eyes.

11.1 UV Equipment Health and Safety

The biological effects of UV radiation result mostly from exposure to wavelengths below 325 nm and resemble the typical syndromes of sunburn, such as skin reddening (erythema), skin burning, dryness, premature aging, and pigmentation. Eyes may be also affected by developing inflammation, pain, photophobia, tearing, temporary blindness, and cataracts.[1] The eyes are most sensitive to UV radiation from 210 nm (UVC and UVB). Maximum absorption by the cornea occurs around 280 nm. Absorption of UVA in the lens may be a factor in producing cataracts (clouding of the lens in the eye). Acute overexposure of the eye to UV radiation can cause photokeratitis (inflammation of the cornea) and photoconjunctivitis (inflammation of the cornea, more commonly known as "snow blindness" or "welders' flash"). Symptoms range from mild irritation to severe pain and, in an extreme case, to irreversible damage.[2]

Industry exposure limits for UV light, established by government agencies, are[1]:

- Near UV (315–400 nm): 1 mW/cm² for exposures longer than 16 min
- Actinic UV (200–315 nm): 0.1 mW/cm² for an 8 h exposure

Hazards due to exposure can be minimized by sound basic engineering, administrative, and hygiene controls. The curing equipment should be properly shielded to prevent escape of UV light into the workplace. Personnel working in the UV curing area should wear special protective eyewear. Access openings to the UV curing equipment should be interlocked to ensure that the radiation source is turned off when the shield is removed. Commercial instruments are available to monitor UV light in the workplace.[2] The Occupational Safety and Health Administration (OSHA) has a standard that covers employee exposure to radiation, *Nonionizing Radiation* (29 CFR 1910.97), and copies of that standard are available from www.osha.gov. Other standards are available from the American Conference of Governmental Industrial Hygienists (ACGIH), which has recommended workplace exposure limits and updates, threshold limit values for chemical substances and physical agents, and biological exposure indices in their ultraviolet radiation section.[2]

The Canadian Centre for Occupational Health and Safety (CCOHS) publishes a web page for comprehensive discussion of guidelines for occupational exposure to UV light and safe practices at www.ccohs.ca/osha answers/phys_agents/ultravioletradiation.html.

Another potential hazard is exposure to ozone, which is generated by most UV lamps, particularly at wavelengths below 200 nm. This colorless gas has a characteristic strong smell and can therefore be detected already at concentrations below its acceptable workplace limit of 0.1 ppm. It will cause biological effects, such as headache, fatigue, dryness of the upper respiratory tract, pulmonary irritation, and may also contribute to respiratory infections.[3] Most UV curing operations use air as the curing atmosphere; consequently, the generation of ozone is a common occurrence. Therefore, the equipment should have ozone filters and sufficient exhaust to prevent escape of ozone and other volatiles into the workplace. The current ACGIH threshold limit value (TLV) ceiling for ozone is 0.1 ppm. The odor threshold concentration is about 0.010.02 ppm, so ozone is detectable by odor before dangerous concentrations are reached.[2] Engineered exhaust systems normally are adequate to prevent escape of ozone into the workplace; however, ozone levels should be monitored regularly to ensure compliance with safe workplace concentrations.[2]

Since the majority of UV lamps generate an appreciable amount of heat, there is a hazard of burning when touching the lamps or certain parts of the equipment. This hazard can be minimized by proper equipment design and by adequate air or water cooling of reflectors, lamp housings, and shutters, and by placing aluminum plate heat sinks below the lamp and the substrate.[4]

Most UV equipment operates at voltages well above the main level, and it is important to prevent exposure to high voltage by proper installation. Microwave-powered lamps have a microwave detector to detect microwave radiation leakage. The system is shut down when microwave irradiance of 5 mW/cm^2 is detected.[4]

With current technology, the use of mercury is crucial in achieving the required light output for practical UV curing lamps and for the necessary lamp efficiency and long lamp life. However, mercury is a toxic element, requiring training in safe use and compliance with strict regulations concerning its disposal. The applicable regulations as well as safe practices of disposal of used lamps are listed in some detail in the literature.[2] The issues of toxicity and high-temperature damage have been eliminated to a high degree by using the ultraviolet light- emitting diodes (UV-LEDs), which do not use mercury and generate much less heat and high temperatures.

11.2 EB Equipment Health and Safety

As mentioned earlier, the stoppage of fast electrons generates X-rays, which are hazardous to human health. X-rays cause cell damage that can lead to cancer formation or genetic mutations. Even at small dosages, X-ray exposure may cause skin burning and general radiation syndrome.

Manufacturers of electron beam equipment provide adequate shielding. The thickness of the shielding and the material used for it depend on the accelerating voltage. Typically, 1 in. (25 mm) thick lead shielding is capable of stopping any X-rays generated by 300 keV accelerators. Most lower-voltage electron beam processors are self-shielded, which means that the electron (and X-ray) source is completely enclosed by shielding. Removable parts of the shielding have been equipped by safety interlocks so that the high voltage of the accelerator is turned off when the shielding is opened. EB units are provided with a radiation detector, which automatically shuts down the power if an alarm setting is exceeded.[1] Units with higher accelerating voltages, typically above 600 keV, require separate vaults constructed from concrete, steel, or a combination of them, enclosing the equipment. State and local governments have regulations controlling the use of radiation-producing equipment, and all electron beam units must be licensed.

The current federal guidelines for occupational exposure to ionizing radiation in the United States are in the Code of Federal Regulations, Part

1910.96. They stipulate that personnel working in the area where ionizing radiation is produced should not receive in any period of a calendar quarter more than the following dose equivalents (rem)[5]:

• Whole body, head and trunk, active blood-forming organs or gonad	1.25
• Hands and forearms	18.75
• Skin or whole body	7.5

Employees working in the area where EB equipment is operating must be monitored for exposure to ionizing (mainly X-ray) radiation with film badges, which detect and quantify any exposure to stray radiation. The employers must train workers thoroughly in the operation of the equipment and proper safety and hygiene, and keep records of exposure of their personnel to ionizing radiation. OSHA has a standard that covers employee exposure to radiation, *Ionizing Radiation* (29 CFR 1910.1096), and copies of that standard are available from www.osha.gov. The regulatory and safety standards for the installation of a low-energy electron beam for commercial use are discussed at some length by Carignano.[6] In the US and Canada, equipment suppliers and end-user locations are required to file with the appropriate state or provincial authorities. The filling is usually submitted to a local Department of Health and Environment. Local radiation control authorities will assist with the procedures to allow the installation and operation to take place with legal compliance. Once the equipment is installed and before put to operation, it must be checked for radiation and interlock safety. After the initial survey and equipment start-up, often the authorities permit an on-site radiation safety officer to take over the responsibility of required quarterly surveys, reports, training, and emergency procedures.

In the European Union, a license is needed to import, install, and operate a self-shielded EB system. The system must meet legal radiation protection and safety rules. It must have a European Conformity certificate, which includes a risk analysis. The entire system and its components must meet CE requirements. There are some specific differences in regulatory and safety standards in individual EU countries. The "Joint-Protocol on Improving Conditions of Use of EB Technology in the Printing and Coating Industry in Europe" published by RedTech Europe covers the main aspects of the industrial use EB systems.

11.3 Chemical Hazards

Materials used in UV/EB curing may present mainly hazards in skin and eye contact and, to a lesser degree, due to accidental ingestion. In some operations, where spray equipment is used, there is some possibility of inhalation of volatilized materials.

TABLE 11.1

Comparison of the Toxicity and Other Properties of UV/EB Curable Systems with Common Solvents

Chemical	Flash Point °F (°C)	VOC	Hazardous Waste	Systemic Skin Irritant	Toxicity	Reproductive Effects
TMPTA[a]	>212 (100)	No	No	Yes	No	No
Oligomer	>>212 (100)	No	No	Maybe	No	No
VM&P naphta	<0 (−17.8)	Yes	Yes	Yes	Yes	No
Toluene	40 (4.4)	Yes	Yes	Yes	Yes	Yes
Xylene	100 (37.8)	Yes	Yes	Yes	Yes	Yes
1-Butanol	100 (37.8)	Yes	Yes	Yes	Yes	Yes

[a] TMPTA = Trimethylolpropane triacrylate.

Acrylate types of monomers and oligomers are known to be skin and eye irritants. Even if they do not cause irritation immediately, they may sensitize a person over a longer exposure and cause allergy.[7] On the other hand, they represent much less of a hazard than common solvents (see Table 11.1).

Some time ago, there has been an issue with the detection of 4-methylbenzophenone (MBP) in breakfast cereal in Europe, leading to publicity concerning the presence of this photoinitiator in the food. Based on the close similarity between MBP and benzophenone, a photoinitiator for inks and coatings that are cured with UV light, the European Food Safety Authority (EFSA) has concluded that short-term consumption of breakfast cereals containing ppm traces of MBP should not pose a risk to most people but recommended that this substance undergo more detailed evaluation if it will continue to be used in food packaging.[8]

There are many aspects of protection of personnel handling the chemicals in a production setting. First, they must be informed about the hazards involved by reading thoroughly the material safety data sheet (MSDS) for each material used. These must be supplied by the manufacturer or vendor of the chemical and must be accessible to the personnel at all times. MSDSs not only inform about the hazards but also recommend proper handling, the type of personal protection to be used, ways of disposal, ways to handle spills, and emergency contacts.

Skin should be protected by gloves (preferably nitrile or butyl rubber), boots, and long-sleeve clothing. For the protection of eyes, safety glasses with side shields or splash goggles are the best choice when handling liquid chemicals. If spray equipment is used, proper respiratory protection may be needed. If any contact with the chemical occurs, the affected area has to be washed with soap and water immediately; solvents should not be used. In extreme cases, proper medical help may be required.

In the European Union, a very comprehensive initiative with the aim to improve the protection of human health and the environment from the hazards of chemicals, and to enhance the competitiveness of the EU's chemical industry, is underway. It is referred to as REACH, which stands for Registration, Evaluation, Authorisation and Restriction of Chemicals. REACH places greater responsibility on industry to manage the risk of chemicals and provide appropriate safety information to professional users and, as far as the most hazardous substances, to consumers. New substances need to be registered before they are placed on the market.[9]

References

1. Golden, R., *RadTech Report*, 11, No. 3, May/June, p. 13 (1997).
2. The Association for UV&EB Technology, *Ultraviolet Curing Lamp Safety and Handling*, RadTech, Bethesda, MD (2010).
3. *UV/EB Curing Primer 1*, 4th ed., RadTech International North America, Northbrook, IL, p. 55 (1995).
4. Mehnert, R., Pincus, A., Janorsky, I., Stowe, R., and Berejka, A., *UV & EB Technology and Equipment*, Vol. 1, John Wiley & Sons Ltd., Chichester/ SITA Technology Ltd., London, p. 269 (1998).
5. *UV/EB Curing Primer 1*, 4th ed., RadTech International North America, Northbrook, IL, p. 57 (1995).
6. Carignano, T., "What are the Basic Regulatory and Safety Standards for the Installation of a Low-Energy Electron Beam for Commercial Use?", *UV + EB Technology*, Vol. 4, No. 4, p. 10 (2018).
7. Bean, A. J., and Cortese, J., *FLEXO*, 25, No. 7, July, p. 37 (2000).
8. *4-Methylbenzophenone*, RadTech briefing, update March 13 (2009).
9. European Commission, Joint Research Centre, Institute for Health and Consumer Protection, June (2009).

12

Recent Developments and Trends

The UV/EB industry is an important part of global economy. Total volume of formulated product that represent this industry is estimated to be 630 kT[1]. The volumes in kT and growth rates (CAGR in %) for individual major regions of the world are shown in Table 12.1:

The *technology split* of the current state of the industry is shown below[1]:

100% Traditional UV/EB	40%
Solvent-borne	32%
Water-based	17%
UV Powder	11%

The application split and growth (CAGR) of the current state of the industry is shown below Table 12.2[1]:

The key industry drivers are based mainly on cost-effectiveness, cost in use, physical properties, productivity reduced energy consumption, and environmental footprint.[1]

The main components of economy where UV/EB technology has been growing are:

- Vehicle building (automotive plastics and composites, wiring and cable coatings, and paints, electronics, adhesives, and fuel cells) and automotive refinishing (see also Section 12.2.10)
- Aerospace (wiring and cables, composites, adhesives, repairs, and marking)
- Housing and construction (wood structures, flooring, furniture, laminates, composites, displays, wiring and cables, and optical fibers)
- Electronics and electrical (wires and cables, shrinkable tubing, displays, conformal coatings, adhesives, and battery separators)
- Medical (coatings, adhesives, sensors and probes, prosthetics and biomedical products, and optical fibers)
- Consumer products (packaging, shrinkable films and sheets, labels, displays, and CDs and DVDs)
- Other (protective coatings, adhesives, shrinkable tubings, displays, and composites)

TABLE 12.1

Volumes and Growth Rates of Individual Regions

USA, kT		Europe, kT		Asia, kT	
2016	123	2016	171	2016	279
2019	134	2019	189	2019	330
CAGR, %	3.0	CAGR, %	2.1	CAGR,%	3.6

TABLE 12.2

Application Split and Growth Rates

Application	%	CAGR, %
Wood coatings	39	3
Printing inks	19	4
OPV[a]	18	3
Adhesives	5	4
Electronics	4	1
Additive manufacturing	3	10
Metal coatings	2	4

[a] Organic photovoltaics.

12.1 Current Trends in Equipment and Chemistry

The main drivers for both equipment and chemistry are cost reduction, energy reduction, and environmental issues (sustainability, "green" raw materials, reduced volatile organic compound (VOC), reduced toxicity, lowering odor, reduced defects, and manufacturing scrap). Companies are increasingly considering sustainability in their business decisions. UV/electron beam (EB) technology offers several sustainability features when compared to thermal curing:

- Reduced use of solvents or their complete elimination
- Reduced usage of energy
- Safer workplace
- Reduced usage of fossil fuels and reduced greenhouse gas emissions
- Reduced or entirely eliminated pollution controls
- Reduced transportation costs due to reduced or eliminated consumption of solvents
- Recyclable inks, coatings, and product wastes

These features have a positive influence on performance and economic returns.

NOTE: The World Commission on Environment and Development defines *sustainability* as "development that meets the need of the present without compromising the ability of future generations to meet their needs."

12.1.1 UV/EB Equipment

The sizes of equipment and its components continue to be reduced to become less costly and consume less energy. Tendencies are to make them portable and modular where possible. Other trends are:

- Growth of the use of robotics
- Growth of 3D curing (both UV and EB)
- Enhancing safety

12.1.2 Chemistry

In chemistry, the main efforts are geared toward improvement of the environmental impact by reducing VOC and using either aqueous resins or resins without any VOC or water. Automotive OEM production and the refinishing business are demanding a highly weather- and scratch-resistant topcoat with no or minimal possible yellowing, improved bonding between individual coatings and paint, as well as dependable dual-cure systems for shadowed areas. In general, further development of fiber-reinforced polymeric composites and gel coats (both UV and EB) is underway for many industrial and consumer products. New monomers and oligomers based on natural and sustainable raw materials are being developed. Relatively numerous applications of nanotechnology are emerging. The rapid growth of the use of UV-LED curing demands new types of photoinitiators, oligomers, and monomers for the cure with new LED lamps.

12.2 Emerging Technologies and Applications

Below are some novel developments and applications. Because the number of them grows at a great rate, this is the only "snapshot" taken at this writing and the list is far from being complete.

12.2.1 The Use of UV-LED in 3D Printing

3D printing is the building of an object by adding successive layers of the desired starting material. The data source for the information is usually a computer file. The different layers can be reacted in place or the physical form changed to create the object. Another name for 3D printing is additive manufacturing—the reason for which is understood, given the definition of 3D printing mentioned above. While there are no hard-and-fast rules, 3D printing usually refers to the making of single items and prototypes, while additive manufacturing tends to refer to the production of many items on a commercial scale. In general, a rapid growth is expected for 3D printing. It is driven mainly by automotive and healthcare. The global market is estimated to be $9.9 billion in 2018 and is expected to grow to $34.8 by 2024.[2]

In the process using a UV-LED lamp, the light from the flat LED panels (arrays) shines directly, in a parallel fashion, onto the build area. The materials used are essentially liquid resins, such as photopolymer solutions.[1] The history of the stereolithography is presented by Ren.[3] A rather thorough study regarding the formulations for 3D printing is reported by Viereckl.[4]

12.2.2 Modification of Polymer Substrates Using Electron Beam-Induced Graft Copolymerization

Electron beams (EBs) are a form of ionizing radiation that have enough energy to break chemical bonds of organic materials, including polymers. The most common result of the irradiation of polymeric materials to EB is the formation of free radicals. The possible reactions following this breakdown are curing (cross-linking), scission (main chain scission), or grafting.

The grafting reaction (copolymerization) occurs when radicals formed in and on polymer substrate become a site for initiation of monomer polymerization. The net result is that two dissimilar polymers are joined to form a new copolymer material. EB grafting is not known as much as cross-linking, but it is an important process for producing new functional materals.[5] The resulting materials depend on the initial materials and monomers used, which include specialty fabrics, reinforcing fiber for composites, plastic films with enhanced adhesion, and media used for separation and purification processes.[5]

12.2.3 Improving Surface Cure with UVC LED Lamps

When using free radical formulation curing with UV-LED lamps, the cured materials are susceptible to oxygen exhibition, which results in the surface of the cured material remaining tacky, even when the layers underneath may be fully cured. There are a number of methods to eliminate this, such as using an inert gas (nitrogen, carbon dioxide) to replace the oxygen on the surface. This method is quite effective but expensive and not convenient. Another common method is to provide sufficient energy in the UVC wavelength range (200–280 nm) to cure the surface of the material. A test program has established that when a UVA LED system is paired with a UVC LED system, providing just a little UVC exposure for post-cure not only provides a tack-free surface but also reduces the total dose requirement.[6]

12.2.4 Applications for Sealed Tube EB Lamps

Low-energy self-shielded EB systems have been used in industrial applications for more than 30 years. Early systems operated mostly in the 175–300 kV range. A new generation of smaller systems operating in the range 80–125 kV was introduced in the early 1990s. In both these systems, electrons are accelerated through metallic window foils in the vacuum environment. Later, a new concept of a small accelerator was designed that was essentially an emitter, which used sealed, thinner metallic foil and maintained permanent vacuum. This concept was more like a lamp. It was a "plug-in" system typically 250–400 mm (10–16 in.) wide. It was used for cross-linking and sterilization. It was easy to handle and operate and offered the possibility to be combined to irradiate wide webs or irradiate in different positions, or in a 3D fashion. The inventor and manufacturer went out of business after several years and eventually sold the technology to a German company as well as to a Japanese company, and it is known that both sell the product. In the meantime, several other manufacturers developed and now sell similar products. At the moment it appears that this concept will serve as a laboratory system and for in-line package sterilization. Another possibility is irradiation of three-dimensional parts.[7]

12.2.5 Advances in Energy Cure Inkjet Printing

Inkjet printing using energy cure (EC) inks is expanding rapidly through the graphic arts and industrial markets. More than 40 companies manufacture UV-cure inkjet ink, about half in North America and several nurturing EB as well. Growth is double digit and is projected to stay at that rate for some years.[8]

Mercury arc UV lamps still dominate the installed base. However, UV-LED is now very popular, when the equipment is being replaced and is preferred, unless it does not perform. Also, the EB cure is becoming a reality. The former enables printing on temperature-sensitive substrates and the latter for the same reason. EB cure is mainly attractive where required cure-through of 1 mm (40 mils) is necessary and when residues from photoinitiators cannot be accepted, such as is the case in food packaging (also see Section 12.2.8). 3D and additive manufacturing represent a relatively small but high-value market.

Packaging offers many applications for cartons, boxes, bags, cans, bottles, and pouches. Surfaces are flat and curved, and there are many opportunities other than food. Electron beam will ensure penetration of the near food market packaging and counteract the migration perception of UV, with the opportunity for large inroads into flexible packaging.[8]

12.2.6 Current State of and Trends in the UV-LED Technology

UV-LED technology has developed rapidly over the past 15 years, mainly in terms of performance, reliability, service life, and output of the light sources. At the same time, processed formulations have been optimized for

UV-LED output. Gradually, UV-LED technology has progressed from UV digital inkjet pinning and spot cure into more demanding, higher speed, and wider commercial application. At the time of this writing, UV-LED curing technology is used in the following applications[9-11]:

- Pressure-sensitive, laminating, structural bonding adhesives
- UV digital inkjet printing, screen, flexo, sheetfed offset, web offset printing methods
- Fiber-optic coating
- Wood coating
- Product assembly, including electronic display and medical devices
- Field repair

Current commercial wavelengths include 365, 385, 395, and 405 nm. For most ink applications, 395 nm wavelength is preferred, with 365 and 385 nm used to a lesser degree. For adhesives, the most widely used wavelengths are 365 and 405 nm, depending on formulation, although wavelengths of 385 or 395 nm work well in some cases. Overprint varnishes tend to match the ink wavelength of 395 nm. There is not yet a consensus about the optimum wavelengths for both functional and hard industrial coating—there is no consensus because the development still is in its infancy.[9] The continuing success in the development of the technology depends on properly identifying the process variables and matching the UV-LED system to the application, formulation, and material-handling equipment.[9]

12.2.7 Advances in Energy-Curable Pressure-Sensitive Adhesives

Energy-curable pressure-sensitive adhesives (EC-PSAs) eliminate the need for drying, solvent extraction, or preheating steps, compared to traditional waterborne, solvent-borne, or hot melt adhesives. These processing benefits make such adhesives particularly suitable for temperature-sensitive substrates and in-line application.

The main components of energy-curable PSAs are[12]:

- *Oligomers* provide much of the shear strength but also affect tack, peel, reactivity, creep resistance, heat resistance, and chemical resistance.[13]
- *Monomers* mainly serve as reactive diluents lowering the viscosity of the formulation. They also control surface wetting, leveling and other physical properties.

- *Photoinitiators* absorb UV energy to produce free radicals that induce polymerization. Formulations for electron beam curing do not need photoinitiators to polymerize.
- *Additives* (stabilizers, tackifiers, adhesion promoters, antioxidants, fillers, plasticizers, pigments, wetting agents and other substances as needed for the given application).

With the wealth of available chemistries, the range of EC-PSA performance is quite broad, and the ability to tailor each formulation to specific performance targets makes this approach very versatile. One interesting advantage that EC-PSAs may be able to offer is low tack with high peel strength, which would potentially be useful for applications where minimal pressure is required for bonding.[12]

12.2.8 EB Curing Technology in Web Package Printing

Electron beam technology is well-established in several web package printing applications. The equipment used for that purpose is known as *low-energy* systems, which the industry defines as systems operating with accelerating voltages between 70 and 300 kV. These systems are completely self-shielding and capable of handling webs with widths from 400 to 2,700 mm (1.3 to 8.9 ft).[14] As pointed out earlier, the penetration of electron beams into the substrate is controlled by the accelerating voltage. At the lower end of this voltage spectrum (70–110 kV) energy can be concentrated in the first 5–20 μm (0.2–0.8 mils), which is ideal for curing inks and coatings in the printing processes.[14] The electron beam energy generates free radicals that initiate polymerization of monomers and oligomers in the ink and coating formulations without the need of photoinitiators, and the consistent cure process makes EB energy attractive for food, pharmaceutical, or personal cure products.[14]

Currently, EB cure is used for web offset package printing. The best-established application in this area is folding cartons, including milk, juice and ice cream, printed on polyethylene (PE) coated paperboard. Other applications of offset package printing include dry food packets, labels, and multiwall bags.[14] EB ink systems have also been developed for flexographic printing. The new ink technologies are optimized so that individual colors may be wet-trapped without interstation driers.[14]

EB curing of inkjet printing is suitable for packaging and label markets[15] (see also Section 12.2.5). Digital printing with EB curing offers the following advantages[15]:

- High quality
- Cost effectiveness for just-in-time (JIT) printing

- Better properties
- Possible for in-line coating and laminating
- Possibility of the use for food

Typical operating conditions for digital printing applications are[15]:

- Web widths 15–30 in. (380–760 mm)
- Maximum speed 300 ft/min (91 m/min)

New applications in development stage include digital printing and narrow web systems using compact sealed tube EB "lamps"[14] (see Sections 4.1.4 and 12.2.4).

12.2.9 Polymer Optics from UV-Curable Polymer Systems

Polymer optics offer several advantages over traditional glass optics including lighter weight, ease of manufacturing, and lower material cost. Because of that, optical glass materials have been successfully replaced with polymer optics in many applications such as eyeglasses, phone screens, and cameras. Current drawbacks of polymer optics include the lack of a high- precision fabrication method as well as limited materials that exhibit equivalent material properties as those of glass optics. This study explored a variety of materials with the goal to create polymer optical materials for direct view sighting systems due to their lighter weights, lower material costs, and ease of manufacturing.

The UV-curable chemistries used in the experiment were thiol-ene and thiol-acrylate systems and were found to produce good quality optics; however, additional progress is needed to fabricate high-quality optical polymers.[16]

12.2.10 UV-LED for Automotive and Transportation Applications

Conventional UV curing using mercury arc systems has been used in the automotive and transportation industries for decades. UV-LED curing is relatively new and has not gained much traction in that area.[17] However, the automotive and transportation industry is facing many challenges in design, engineering, and manufacture. The competitive nature of this industry will require new manufacturing techniques and materials, many of which will be polymers. In fact, a number of UV-LED curing applications are being considered or beginning to be used in this industry. One of the biggest advantages of using UV-LED curing is the reduced heat transfer to the parts and substrates when curing inks, coatings, and adhesives. This is clearly very important when the industry is shifting to a large number of different plastic materials, lighter metallic alloys, and fiber-reinforced polymer composites to reduce vehicle weight. Dissimilar and often heat-sensitive materials have to be bonded by using structural adhesives, many of which can be formulated for UV-LED cure. Additional benefits of using UV-LED curing are lower cost of energy and a potential lower amount of scrap due to instant on/off and superior, dependable performance of the UV-LED system.

Based on current interest and development activity, it is anticipated that the following applications will be eventually further developed and expanded[17]:

- Screen-printed in-mold decorated substrates for interior vehicle assemblies
- Structural bonding adhesives
- Conformal coatings
- Sealants, encapsulants and potting of assemblies, wire harnesses, and cables
- Printed appliqués
- Photoresist masks
- Touch-up materials used in automotive refinishing
- Coatings for mirrors, headlights, reflectors, and lenses
- Physical vapor depositions (PVDs) on plastic parts

References

1. Engberg, D., "RadTech Market Survey, 2019," in *RadTech Europe 2019 Conference*, (October 19–20, Munich, Germany) (2019).
2. Brodine, D., "Market Outlook: The Future of the 3D Printing," *UV + EB Technology*, Vol. 5, No. 4, p. 20 (2019).
3. Ren, K., "Stereolithography: Three Decades of UV Technology Innovation," *UV + EB Technology*, Vol. 1, No. 1, p. 12 (2015).
4. Viereckl, J. A., et al., "Formulating for 3D Printing: Constraints and Components for Stereolithography," *UV + EB Technology*, Vol. 4, No. 4, p. 52 (2018).
5. Lapin, S. C., "Modification of Polymer Substrates Using Electron Beam-Induced Copolymerization," *UV + EB Technology*, Vol. 1, No. 1, p. 52 (2015).
6. Kay, M., "Improving Surface Cure with UVC LEDs," *UV + EB Technology*, Vol. 4, No. 1, p. 26 (2018).
7. Lapin, S. C., "What Types of Applications are Enabled by Sealed Tube EB Lamps?", *UV + EB Technology*, Vol. 2, No. 2, p. 12 (2016).
8. Taylor, D., and Cahill, V., "Advances in Energy Cure Inkjet," *UV + EB Technology*, Vol. 3, No. 3, p. 14 (2017).
9. Heathcote, J., "State of the UV-LED Curing Applications," *UV + EB Technology*, Vol. 5, No. 1, p. 18 (2019).

10. Arcenaux, J. A., Wang, T., and Buono, C., "UV LED Cure Applications," in *Big Ideas Conference*, (March 19–20, Redondo Beach, CA) (2019).

11. Kiyoi, E., *The State of UV-LED Curing: Investigation of Chemistry and Applications, UV+EB Technology*, RadTech ebook, RadTech International, NA, p. 60 (2016).

12. Orilall, C., et al., "Recent Progress in UV-/EB-Curable Pressure Sensitive Adhesives," *UV + EB Technology*, Vol. 5, No. 3, p. 46 (2019).

13. Lu, J., and Dong, C., "Radiation Curable Pressure Sensitive Adhesives," in *Proceedings, RadTech Conference 2016*, (May 16–18, Chicago) (2016).

14. Lapin, S. C., "How is EB-Curing Technology Being Used in Web Package Printing Applications?" *UV+ EB Technology*, Vol. 1, No. 2, p. 8 (2015).

15. Rangwalla, I., "New Compact Low Voltage EB Equipment for Curing Inkjet for Packaging and Label Markets," in *Big Ideas Conference*, (March 19–20, Redondo Beach, CA) (2019).

16. Garton, S., et al., "Using UV-Curable Polymer to Create Optics," *UV + EB Technology*, Vol. 4, No. 4, p. 16 (2015).

17. Heathcote, J., "The Growing Viability of UIV LED for Automotive and Transportation Applications," *UV + EB Technology*, Vol. 3, No. 1, p. 34 (2017).

Glossary

abatement (technology): various processes and methods (e.g., incinerators) designed to eliminate or reduce the amount of hazardous waste, environmental emissions, or effluents from a facility

abrasion: the surface loss of a material due to frictional forces

abrasion resistance: reciprocal of abrasion loss

absorbance: an index of the light absorbed by a medium compared to the light transmitted through it. Numerically it is the logarithm of the ratio of incident spectral irradiance to the transmitted spectral irradiance

absorbed dose: the amount of energy absorbed per unit mass of the irradiated material

absorptivity (or absorption coefficient): absorbance per unit thickness of the medium

accelerator: 1. Equipment for the production of high-energy beams of elementary particles, such as electrons or protons, through application of electrical or magnetic forces. 2. Rubber compounding ingredient used in small amounts with a vulcanizing agent to increase the rate of vulcanization (cross-linking) of the base elastomer(s)

addition polymerization: a type of polymerization in which the small molecules (monomers) combine chemically to form polymer molecules without a by-product material being formed

additive lamps: arc or microwave medium-pressure mercury vapor UV lamps that have small amounts of different metal halides added to the mercury in the lamp. The halides emit their characteristic wavelengths in addition to mercury additions. (This term is preferred over *doped lamps*.)

adhesion: the state in which two surfaces are held together by interfacial forces, which may consist of valence forces or interlocking action or both

adhesive: a material that when applied will cause two surfaces in contact with each other to adhere

aging: the irreversible change of material properties after exposure to an environment for an interval of time (also used as the term for exposing the material to an environment for an interval of time)

alpha particle (symbol α): a positively charged particle emitted by certain radioactive materials. It is made up of two neutrons and two protons bound together; thus, it is identical to a helium atom. It is the least penetrating of the three most common types of radiation (alpha, beta, and gamma) emitted by a radioactive material and can be stopped, for example, by a sheet of paper. It is not dangerous to plants, animals, or humans unless the alpha-emitting substance has entered the body

ambient temperature: room temperature or temperature in the surrounding area

amorphous phase: the part of a polymeric material that has no particular ordered arrangement in contrast to the crystalline phase, which is ordered. Semicrystalline polymers consist of different ratios of crystalline and amorphous phases

Angstrom (symbol Å): unit of length used for electromagnetic radiation, atomic, and their components, equaling 10^{-8} cm. It is used very seldom, since it has been replaced by a new unit, nanometer (nm), $1 \text{ nm} = 10^{-9}$ m

anionic polymerization: a process proceeding by the addition of certain monomers to active center-bearing whole or partial negative charges

annealing: a heat treatment process directed at improving performance by removing stresses in the material formed during its fabrication. Typically, the part, sheet, or film is brought to a specific temperature for a definite period of time and then cooled down slowly to ambient temperature

antifoaming agent: a chemical that is added to a liquid mix to prevent formation of foam, or added to the foam itself to break a foam already formed

atomic mass unit: one-twelfth the mass of a neutral atom of the most abundant isotope of carbon, ^{12}C

atomic number (symbol Z): the number of protons in the nucleus of an atom, and also its positive charge. Each chemical element has a characteristic atomic number

atomic weight: the mass of an atom relative to other atoms. The atomic weight of an atom is approximately equal to the total number of protons and neutrons in its nucleus

atomizing: dispersing solid or liquid particles into air

back ionization: excessive buildup of charged particles limiting further deposition onto substrate

backscatter: reflected or scattered radiation in the general direction of the source after it strikes matter (gas, liquid, or solid)

ballast: an inductive transformer device that stabilizes the amount of current flowing through a bulb so that the power remains constant

bandwidth: the range of wavelengths between two identified limits

beam: a stream of particles or electromagnetic radiation going into a single direction

beta particle (symbol β): an elementary particle emitted from a nucleus during radioactive decay, with a single electrical charge and mass equal to 1/1,837 that of a proton. A negatively charged beta particle is identical to an electron. A positively charged beta particle is called a positron. Beta radiation can cause burns, and beta emitters are harmful if they enter the human body. Beta particles are stopped easily by a thin sheet of metal

binning: process of sorting LED chips into groups according to peak irradiance wavelength, tolerance, and forward voltage

betatron: a doughnut-shaped accelerator in which electrons, traveling in an orbit of constant radius, are accelerated by a changing electromagnetic field. Energies up to 340 MeV have been attained in betatrons

blowing agent: compounding ingredient used to produce gas by chemical or thermal action, or both, in manufacture of cellular or hollow articles

borosilicate glass: a strong heat-resistant colorless silica glass that contains a minimum of 5% boric oxide, exhibits exceptional thermal shock resistance, and transmits a greater percentage of ultraviolet energy than glass. It is commonly used as a material for the emitting window on a UV-LED head

bremsstrahlung: electromagnetic radiation emitted by a fast-moving charged particle (usually electron), when it is slowed down (or accelerated) and deflected by the electric field surrounding a positively charged atomic nucleus. X-rays, produced in ordinary X-ray machines, are produced by this mechanism, and so are X-rays produced during irradiation of materials by electron beam. (In German, the term *bremsstrahlung* means "braking radiation.")

bulb: a sealed quartz tube that contains mercury or other fill and is often fitted with electrical connections

butadiene: a gaseous hydrocarbon of the diolefin series. Can be polymerized into polybutadiene or copolymerized with styrene or acrylonitrile to produce SBR and NBR, respectively

cable, electrical: either a stranded conductor with or without insulation, and other coverings (single-conductor cable), or a combination of conductors insulated from one another (multiple-conductor cable)

cable core: the portion of an insulated cable lying under the protective covering or coverings

cable sheath: the protective covering applied to cables

calender: a precision machine equipped with three or more heavy rolls, internally heated or cooled (or both), revolving in opposite directions, used for continuously sheeting and plying up elastomeric compounds, and frictioning or coating with elastomeric compounds

carbon black: finely divided carbon made by incomplete combustion or decomposition of natural gas or petroleum-based oils in different types of equipment. According to the process and raw material used, it can be furnace (e.g., HAF), thermal (e.g., MT), or channel carbon black (e.g., EPC), each having different characteristics, such as particle size, structure, and morphology. The addition of different types of carbon blacks to rubber compounds results in different processing behavior and vulcanizate properties

cathode rays: a stream of electrons emitted by a cathode, or negative electrode, of a gas-discharge tube or by a hot filament in a vacuum tube, such as a television tube

cationic polymerization: a process in which the active end of the growing polymer is a positive ion

Chafer strip: a strip of rubber-coated fabric partially covering the bead assembly of a tire and extending above the rim line. Its function is the chafing of the bead on the rim

channel black: a form of carbon black produced from natural gas by the channel process

chip: a fully functioning minute slice of a semiconductor material, such as silicon, germanium, and gallium arsenide, doped and processed to have p-n characteristics. Specifically, gallium nitride (GaN) is used to generate longer ultraviolet and blue visible wavelengths. In referring to LED, the term chip is often used interchangeably with *diode*, die, and semiconductor. See also **Die**

chlorosulfonated polyethylene: a product obtained by treatment of polyethylene by chlorine and sulfur dioxide. It is an elastomer highly resistant to chemicals and ozone

cold mirror: a reflector that is coated by a dichroic material that absorbs or passes wavelengths in the infrared region while reflecting those in the UV range

composite: in polymer technology, a combination of a polymeric matrix and a reinforcing fiber with properties that the component materials do not have. The most common matrix resins are unsaturated thermosetting polyesters and epoxies, and reinforcing fibers are glass, carbon, and aramid fibers. The reinforcing fibers may be continuous or discontinuous. Some matrix resins are thermoplastics

compound (elastomeric): an intimate mixture of a polymer (or polymers) with all the ingredients necessary for the finished article

compression molding: a fabrication method in which a polymeric material, mostly a thermoset (a plastic or an elastomer), is compressed in a heated mold for a specific period of time

compression set: the residual deformation of a material after removal of the compressive stress

COPA: copolyamide (thermoplastic elastomer)

COPE: copolyester (thermoplastic elastomer)

copolymer: a polymeric material formed by the reaction of two or more monomers

crystallinity, crystallization: orientation of the disordered long-chain molecules into repeated patterns

cure (curing): 1. Polymerization or cross-linking (or both) by radiation (UV or EB). 2. Cross-linking or vulcanization of elastomeric materials

cyclotron: a particle accelerator in which charged particles receive repeated synchronized accelerations by an electrical field as the particles spiral outward from their source. The particles are kept in the spiral by a powerful magnetic field

damping: the dissipation of energy with time or distance

degradation: deterioration, usually in the sense of a physical or chemical process, rather than a mechanical one

dichroic: exhibiting significantly different reflection or transmission in two different wavelength ranges

dichroic reflector: a reflector having reduced reflectance to long wavelengths (IR radiation); also called "cold mirror"

die: a fully functioning, minute slice of a semiconducting material, such as silicon, germanium, and gallium arsenide, doped and processed to have p-n junction characteristics. Specifically, gallium nitride (GaN) is used to generate longer ultraviolet and blue visible wavelengths. In referring to LED, the term *chip* is often used interchangeably with *diode*, die, and semiconductor. See also **Chip**

dielectric constant: that property of a dielectric (insulating material) that determines the electrostatic energy stored per unit volume for unit potential gradient

dielectric loss angle (symbol δ): an angle between the vector for the amplitude of the total current and that for the amplitude of the charging current. The tangent of this angle (tan δ) is the loss tangent, a direct measure of the dielectric loss

dielectric strength: the voltage that an insulating material can withstand before breakdown occurs, usually expressed as voltage gradient (volts per mil or mm)

diffuse: a characteristic of a surface that reflects or scatters light in all directions

diode: a common electrical device that is added to a circuit as a means of restricting the flow of electricity; it can be thought of as a switch or a valve. Its key property is that it conducts electricity in only one direction. Also see **p-n junction**

doped lamp: term applied to a UV mercury lamp containing metal halide added to the mercury to alter the emission spectrum of the lamp (preferred term is *additive lamp*)

doped (LED): an LED semiconductor material that has been impregnated with impurities to produce a specific n-type or p-type conductivity

dose: 1. In EB processing: the amount of energy absorbed per unit mass, unit 1 Gray (Gy) = 1 J/kg. 2. In UV processing: a common but loosely used term for *irradiant energy density* or *flux density* at a surface of the medium of interest (unit J/cm^2)

driver board: a printed circuit board (PCB) that distributes the DC voltage to the LED chips or modules in an array and provides the pulse width modulation (PWM) capability of the system

durometer: an instrument used for measuring hardness of an elastomeric or plastic material

durometer hardness: an arbitrary numerical value that indicates the resistance to indentation of the indentor point of a durometer

dynamic exposure: exposure to a varying irradiance, such as when a lamp passes over a surface, or a surface passes under a lamp. In such a case, the energy density is the time integral of the irradiance profile

dynamic properties: mechanical properties of polymeric materials exhibited under repeated cyclic deformation

effective irradiance: radiant power, within a specified wavelength range, arriving at a surface per unit area (unit J/cm^2). If the wavelength range is understood, the term is shortened to *irradiance*

Einstein: one mole of photons

elasticity: the property of a matter by virtue of which it tends to return to its original size and shape after removal of the stress causing deformation, such as stretching, compression, or torsion

elastomer: a macromolecular (polymeric) material that, at room temperature, is capable of recovering substantially in shape and size after removal of a deforming force

electrodeless: referring to microwave-powered UV systems

electroluminescence: an optical and electrical phenomenon inherent to LEDs in which a material emits light energy when an electric current is passed through it

electromagnetic radiation: radiation consisting of associated and interacting electric and magnetic waves that travel at the speed of light (e.g., light, radio waves, gamma rays, X-rays); all can be transmitted through a vacuum

electromagnetic spectrum: the full wavelength range of electromagnetic radiation, including microwave, ultraviolet, visible, and infrared energy

electron: an elementary particle with a unit negative electric charge and a mass of 1/1,837 that of a proton. Electrons surround the positively charged nucleus of an atom and determine the chemical properties of the atom

electronvolt (eV): the amount of kinetic energy gained by an electron when it is accelerated through an electric potential difference of 1 volt. It is equivalent to 1.603×10^{-12} erg. It is a unit of energy, or work, not of voltage

elongation: extension of a body produced by a tensile stress

elongation, ultimate: the elongation at the time of rupture

emission spectrum: radiation from an atom in an excited state, usually displayed as radiant power versus wavelength. Each atom or molecule has a unique spectrum. The spectra can be observed as narrow line emission (atomic emission spectra) or as quasi-continuous emissions (molecular emission spectra). A mercury plasma emits both line spectra and continuous spectra simultaneously

emitting window: flat, rectangular piece of quartz or borosilicate typically secured at the base of an LED head to protect the dies while simultaneously transmitting ultraviolet wavelength

energy density: amount of radiant energy arriving at a surface per unit of area, commonly expressed in J/cm^2 or mJ/cm^2. It is an integral of irradiance over time

extrusion: a process in which heated or unheated polymeric material (plastic or elastomer) is forced through a shaping orifice (die) in one continuous shape, as in film, sheet, slab, profile, pipe, coating, etc

fatigue, dynamic: the deterioration of a material by a repeated deformation

filament winding: a process to produce a composite part from a continuous fiber impregnated by a matrix resin by winding it onto a rotating or stationary mandrel

fluence: the time integral of fluence rate (J/m^2 or J/cm^2). For a parallel and perpendicularly incident beam, not scattered or reflected, *energy density* and *fluence* become identical

fluidization: the process of suspending powder particles using compressed air, creating a fluid mixture of air and powder

flux (radiant flux): the flow of photons (in Einstein/second)

forward voltage: the actual voltage across a semiconductor diode carrying a forward current

free radical: a reactive species having an unpaired electron that initiates a reaction with a double bond, for example, in acrylate polymerization. It is produced from its stable paired state by energy absorption

free radical polymerization: a process with a complex mechanism of initiation, propagation, and termination, of which the propagation and termination steps are typically very fast

frequency: the number of times a periodic wavelength cycle occurs in 1 s; the unit of measurement is hertz (Hz)

frictioning: process of impregnating woven fabric with rubber compound using a calender whose rolls rotate at different surface speeds

FTPE: fluorinated thermoplastic elastomers, thermoplastic elastomers containing mainly fluorine or fluorine and chlorine (in some cases)

furnace black: a carbon black obtained by incomplete burning of natural gas or petroleum oil, or both in a large furnace

gamma rays (symbol γ): high-energy, short-wavelength electromagnetic radiation. Gamma radiation frequently accompanies alpha and beta emissions and always accompanies nuclear fission. Gamma rays are very penetrating and are best stopped or shielded against by dense materials, such as lead or depleted uranium. Gamma rays are similar to X-rays, but are usually more energetic and are nuclear in origin

gel: a semisolid system consisting of a network of solid aggregates in which liquid is held

glass transition (temperature), T_g: the temperature at which the amorphous portion of a semicrystalline solid changes from its glassy state, becoming soft and flexible (rubbery); not to be confused with melting temperature

graft copolymer: a copolymer in which chains of one polymer are attached to chains of a previously formed polymer or copolymer in such a way that the junction points have three or more chains attached

gray (Gy): SI unit of radiation-absorbed dose, equals 1 J/kg; it has replaced the older unit, *rad*, which is 100 ergs/g (or 0.01 J/kg)

ground state: the state of nucleus, atom, molecule, or any other particle at its lowest (normal) energy level

gum (compound): unfilled elastomeric compound containing only ingredients necessary for a sufficient cross-linking

HAF (high abrasion furnace black): highly reinforcing furnace carbon black increasing resistance of a rubber compound to abrasion

heat aging: exposure of polymeric materials under specified conditions (temperature, time, presence or absence of air or oxygen, etc.), then testing them in stress-strain and hardness, determining the change of properties in comparison to the original (unaged materials)

hysteresis: energy loss, the difference between the work output in a cycle of extension and retraction

impact resistance: resistance to fracture under shock force

intensity: a generic term, with a variety of meanings, frequently used to mean *irradiance*, which is the correct and preferred term

interpenetrating network (IPN): a combination of two polymers into a stable interpenetrating network. In a true IPN, each polymer is cross-linked to itself but not to the other, and the two polymers interpenetrate each other. In a semi-IPN, only one of the polymers is cross-linked; the other is linear and by itself would be a thermoplastic. The purpose of producing IPN is to improve strength, stiffness, and chemical resistance of certain polymeric systems

ionization: the process of adding one or more electrons to, or removing one or more electrons from, atoms or molecules, thereby creating ions. High temperatures, electric discharges, nuclear radiation, or high-energy electrons can cause ionization

ionizing radiation: any radiation displacing electrons from atoms or molecules (e.g., alpha, beta, and gamma radiation). Ionizing radiation may produce severe skin and tissue damage

irradiance: radiant power arriving at a surface from all forward angles, per unit area, expressed in watts per unit area (W/cm^2 or mW/cm^2)

irradiance profile: the irradiance pattern of a lamp or, in the case of dynamic exposure, the varying irradiance at a point on a surface that passes through the field of illumination of a lamp or lamps; irradiance versus time

irradiation: application of radiation to an object

irradiator (lamp head): assembly or fixture that is a sheet metal housing containing the UV bulb, reflector, shutter, magnetrons, or screens; it may also include a blower

isotope: one or more atoms with the same atomic number (the same chemical element) but with different atomic weights. The nuclei of isotopes have the same number of protons but a different number of neutrons. Thus, ^{12}C, ^{13}C, and ^{14}C are isotopes of the same element, carbon, and the superscripts denoting their differing mass numbers or approximate atomic weights. Isotopes usually have very nearly the same chemical properties but somewhat different physical properties

Joule: a unit of work or energy (N-m), abbreviated as J. It is a time integral of power

latex: an aqueous colloidal emulsion of an elastomer (natural or synthetic) or a plastic. It generally refers to the emulsion obtained from a tree or plant or produced by emulsion polymerization

LED: light-emitting diode (see below)

LED array: packaged subassembly or module typically consisting of multiple diodes or chips that are individually wire bonded to a printed circuit board and then secured to a heat sink. Also refers to a full curing assembly, which includes numerous modules or LED chips, as well as a cooling fan, or tube fittings, a manifold block, an emitting window, and a sheet metal or plastic outer housing

LED package: an assembly containing one or several chips physically electrically assembled together with the means of electrically connecting the entire assembly to another device

LED irradiator, LED head, or LED light engine: a UV curing assembly, which includes multiple LED chips or modules, a thermal heat sink, a cooling fan or tube fittings, a manifold block, an emitting window, a sheet metal or plastic outer housing, and sometimes, the driver boards

lens: transparent device used to physically protect LED chips, block moisture, and evenly manipulate or spread the emitted UV radiation. It is often made of silicone, borosilicate, or quartz

lethal dose: a dose of ionizing radiation sufficient to cause death. Median lethal dose (MLD or LD50) is the dose required to kill within a specified period of time (usually 30 days) half of the individuals in a large group or organisms similarly exposed. The LD50/30 for humans is 400–500 rads (4–5 Gy)

light-emitting diode (LED): a two-lead semiconductor light source. It is a p–n junction diode that emits light when activated. When a suitable voltage is applied to the leads, electrons are able to recombine with electron holes within the device, releasing energy in the form of photons. Typical materials used for light-emitting diodes are gallium arsenide (GaAs), gallium phosphide (GaP), and aluminum gallium nitride (AlGaN). The exact choice of the semiconductor material used will determine the overall wavelength of the photon light emissions. Also see **Diode** and **p-n junction**

linac: short for linear accelerator

linear accelerator: a long, straight tube (or series of tubes) in which charged particles (ordinarily electrons or protons) gain in energy by the action of oscillating electromagnetic fields

line emission: narrow lines of emission from an atom in the excited state; the "spikes" observed in spectrometry

magnetron: component contained inside a microwave-powered lamp head that converts high-voltage electrical input into microwave energy at 2,450 MHz

MDF: medium-density fiberboard. A composite panel product manufactured from wood fibers and synthetic resin binders bonded together under heat and pressure; the fibers and resin form a homogeneous board with consistent properties in each direction

micrometer (µm): unit of length equal to 1 millionth of a meter; replaces an older term, *micron*

microwave: designating the part of the electromagnetic spectrum associated with the larger infrared waves and the shorter radio waves, between 1 and 10 mm

modulus: the ratio of stress to strain. In the physical testing of rubber, it is the force in pounds per square inch (psi) of original area or pascals (Pa) related to original area to produce a state percentage elongation

monochromatic: light radiated from a source that is concentrated in only a very narrow wavelength range (bandwidth). This may be accomplished either by filters or by narrowband emission

monochromator: an instrument that separates incoming radiant energy into its component wavelengths for measurement

monomer: a low-molecular-mass substance consisting of molecules capable of reacting with like and unlike molecules to form a polymer

nanometer: unit of length commonly used to define wavelength of light, particularly in the ultraviolet and visible ranges of the electromagnetic spectrum. It equals 10^{-9} m or 10^{-3} µm or 10 Å

neutron: an uncharged elementary particle with a mass slightly greater than that of the proton and found in the nucleus of every atom heavier than hydrogen (atomic weight = 1)

ozone (formula O_3): an allotropic form of oxygen produced by the action of electric discharges or of a certain ultraviolet wavelength of light on oxygen. It is a gas with a characteristic odor and is a powerful oxidizing agent

peak irradiance: the intense peak of focused power under a lamp, the maximum point of the irradiance profile. The unit of measurement is W/cm^2

permanent set: the amount by which an elastic material fails to return to its original form after deformation

photon: the carrier of a quantum of electromagnetic energy. Photons have an effective momentum but no mass or electrical charge

phr: abbreviation used in formulation of elastomeric compounds indicating parts in mass units per hundred parts of rubber

Ply: most commonly applies to a layer of rubber-coated fabric (as in tire or belt design). Polychromatic or polychromic consisting of many wavelengths

p-n junction: a p-n junction (positive-negative junction) is a specially engineered diode that is made of many very thin layers of semiconductive materials, where each layer is less than 1 μm thick. Also see **Diode**

polymer: a macromolecular material formed by the chemical combination of monomers having either the same or different chemical composition

polymerization: a chemical reaction in which the molecules of monomers are linked together to form large molecules whose molecular mass is a multiple of that of the original substance. When two or more monomers are involved, the process is called copolymerization or heteropolymerization

power (of a UV lamp): operating power of a tubular lamp is described in watts/inch (W/in) or watts per cm (W/cm). It is derived by simply dividing the electric power by the effective length of the lamp

proton: an elementary particle with a single positive electrical charge and a mass approximately 1,837 times that of an electron; the nucleus of the ordinary, or light hydrogen, atom. Protons are constituents of all nuclei. The atomic number (Z) of an atom is equal to the number of protons in its nucleus

PTFE: polytetrafluoroethylene, also know generally as "Teflon"

quantum yield: a measure of a photochemical efficiency of a photochemical reaction, expressed as the ratio of the number of chemical events per unit time to the number of photons absorbed per unit time; it is a unitless quantity

rad (acronym for *radiation-absorbed dose*): the original basic unit of absorbed dose of ionizing radiation; one rad equals 100 ergs of radiation energy per gram of absorbing material. This unit has been replaced by the gray (Gy)

radiachromic: exhibiting a change of color or optical density with exposure to UV or EB radiation. These changes can be correlated to the amount of exposure

radiant energy: energy transfer, expressed in joules or watt-seconds ($1 J = 1 W$)

radiant power: rate of energy transfer, expressed in watts or joules per second ($1 W = 1 J/z$)

radiation illness: an acute organic disorder that follows exposure to relatively severe doses of ionizing radiation. It is characterized by nausea, vomiting, diarrhea, blood cell changes, and in later stages, hemorrhage and loss of hair

radiometer: an instrument that senses irradiance incident on its sensor element and may incorporate either a thermal or photonic detector

rem (acronym for *roentgen-equivalent man*): the unit of dose of ionizing radiation that produces the same biological effect as a unit of absorbed dose of ordinary X-rays

resilience: the ratio of energy output to energy input in a rapid (or instantaneous) full recovery of a deformed specimen

responsivity (spectral sensitivity): the response or sensitivity of any system in terms of incident wavelength. In radiometry, it is the output of a device versus wavelength

RF (radio frequency): any frequency between normally audible sound waves and the infrared portion of the spectrum lying between 10 kHz and 10^6 MHz

RoHS (Restriction of Hazardous Substances) Directive: restricts the use of certain dangerous substances in electrical and electronic equipment

rubber: a material that is capable of recovering from large deformations quickly and forcibly and can be, or already is, modified to a state in which it is essentially insoluble (but can swell) in boiling solvent such as benzene, methyl ethyl ketone, and ethanol-toluene azeotrope. A rubber in its modified state, free of diluents, retracts within 1 min to less than 1.5 times of its original length after being stretched at room temperature to twice its length and held for 1 min before release

SBS: styrene-butadiene-styrene block copolymer

SEBS: styrene-ethylene-butylene-styrene block copolymer

SEPS: styrene-ethylene-propylene-styrene block copolymer

semicrystalline polymer: a material consisting of a combination of crystalline and amorphous regions. Essentially, all common plastics and elastomers with the tendency to crystallize are semicrystalline. The degree of crystallization depends on the structure of the polymer and the conditions of fabrication

spectral absorbance (absorbance spectrum): absorbance described as a function of wavelength

spectral irradiance: irradiance at a given wavelength per unit wavelength integral, expressed in $W/cm^2/nm$; measured usually with a spectroradiometer

spectral output: the radiant output of a lamp versus wavelength. It is displayed in a variety of ways, the most common being the integration of the energy over 10 nm bands, which reduces the difficulty of quantifying the effects of line emission spectra

spectroradiometer: an instrument that combines the functions of a radiometer and a monochromator to measure irradiance in finely divided bands

static exposure: exposure to a constant irradiance for a controlled period of time

sulfur: chief vulcanizing agent for many elastomers, particularly those based entirely or part on butadiene or isoprene

tack (or building tack): a property of an elastomer or rubber compounds that causes two layers of compound that have been pressed together to adhere firmly in the areas of contact. It is very important for building tires or other laminated structures

target: material subjected to particle bombardment (as in accelerator) or irradiation (as in reactor) in order to induce nuclear reaction

tentering: a continuous process for drying cloth while held under tension to remove wrinkles and give a smooth surface, also for orienting polymeric films or sheets in the transverse (cross-machine) direction

tentering frame: a machine used to stretch textiles and polymeric films and sheets mostly for directional orientation

terpolymer: a copolymer made from three different polymers

thermal black: soft carbon black formed by the decomposition of natural gas (e.g., MT, medium thermal black). It has little stiffening effect but imparts toughness, resilience, good resistance to tearing, and fair abrasion resistance

thermoplastic: a material capable of being repeatedly softened by increase of temperature and hardened by decrease of temperature (*see* **Plastics**)

thermoplastic elastomer (TPE): a polymeric material that is elastic at ambient or moderately elevated or lowered temperature that can be processed and recycled as a thermoplastic (i.e., by melt processing). The processing and service temperatures depend on the chemical nature of the material

tire, pneumatic: a tire casing (consisting of cord fabric), tread, sidewalls, and bead with or without an inner tube, capable of being inflated with a gas

TPU: thermoplastic polyurethane (thermoplastic elastomer)

vulcanization: an irreversible process during which an elastomeric compound, through a change in its chemical structure (e.g., crosslinking), becomes more elastic and more resistant to swelling by organic liquids, and elastic properties are conferred, improved, or extended over a greater range of temperature

Watt: the unit of power equal to the work done at the rate of 1 joule per second, or the power produced by a current of 1 ampere across a potential difference of 1 volt; 1/746 horsepower; abbreviated W, a smaller unit is a *milliwatt* (mW). In optics, it is a measure of radiant or irradiant power

waveguide: a device directing microwaves toward the bulb in microwave-powered UV systems

wavelength: a fundamental descriptor of electromagnetic energy, including light. It is the distance between corresponding points of a propagated wave, frequently using the symbol λ (lambda). It is the velocity of light divided by the equivalent frequency of oscillation associated with a photon. The unit of measurement is a *nanometer* (10^{-9} m)

X-ray: a penetrating electromagnetic radiation emitted when the inner orbital electrons of an excited atom return to their normal state. X-rays are usually nonnuclear in origin and are generated by bombarding a metallic target with high-speed electrons

Z: the symbol for atomic number

Appendix I: Bibliography

Below is a list of books that may provide additional and often detailed information on various topics in radiation science and technology. They are listed chronologically, starting with the most recent publications. Efforts were made to make this list as complete as possible.

Kumar, V., Chaudhary, B., Dharma, V., and Verma, V., (Eds.), *Radiation Effects in Polymeric Materials*, Springer Nature, Switzerland (2019).

Drobny, J.G., *Ionizing Radiation and Polymers-Principles, Technology and Applications*, Elsevier, Oxford, UK (2013).

Makuuchi, K., and Cheng, S., *Radiation Processing of Polymer Materials and Its Industrial Applications*, John Wiley & Sons, Hoboken, NJ (2012).

Greene, W. A., *Industrial Photoinitiators, Technical Guide*, CRC Press, Boca Raton, FL (2010).

Drobny, J. G., *Radiation Technology for Polymers*, 2nd ed., CRC Press, Boca Raton, FL (2010).

Glöckner, P., Jung, T., Struck, S., and Studer, K., *Radiation Curing: Coatings and Printing Inks (European Tech Files)*, Vincentz Network, Hannover (2008).

Schwalm, R., *UV Coatings, Basics, Recent Developments and New Applications*, Elsevier, Amsterdam (2007).

UV Technology: Practical Guide for All Printing Processes, Berufgennosenschaft Druckund Papierverarbeitung, Wiesbaden, Germany (2007).

RadTech Printer's Guide, RadTech International North America, Bethesda, MD, www.radtech.org.

Läuppi, U. V., EB/UV/γ terms *(English-German and German-English Dictionary)*, Vincentz Network, Hannover (2003).

Drobny, J. G., *Radiation Technology for Polymers*, CRC Press, Boca Raton, FL (2003).

Talbert, R. (Ed.), *UV Powder Coating Application Guide*, RadTech International North America, Chevy Chase, MD (2002), www.radtech.org.

Koleske, J. V., *Radiation Curing of Coatings*, ASTM International, West Conshohocken, PA (2002).

Dietliker, K., *Compilation of Initiators Commercially Available for UV Today*, SITA Technology Ltd., Edinburgh and London (2002).

ASTM Standards Related to Testing of Radiation-Cured Coatings, ASTM International, Conshohocken, PA (2001), www.astm.org.

Davidson, R. S., *Exploring the Science, Technology and Applications of U.V. and E.B.*, SITA Technology Ltd., London (1999).

Crivello, J., and Dietliker K., *Photoinitiators for Free Radical Cationic & Anionic Photopolymerization*, 2nd ed., John Wiley & Sons, New York (1998).

Mehnert, R., Pincus, A., Janorsky, I., Stowe, R., and Berejka, A., *UV & EB Technology & Equipment*, Vol. I., John Wiley & Sons Ltd., Chichester and SITA Technology, Ltd., London (1998).

Clough, R. L., and Shallaby, S. W. (Eds.), Irradiation of Polymers, Fundamentals and Technological Applications, *ACS Symposium Series* 620, American Chemical Society, Washington, DC (1996).

Garrat, P. G., *Strahlenhärtung*, Curt R. Vincentz Verlag, Hannover, Germany (1996).

Fouassier, J.-P., *Photoinitiation, Photopolymerization, and Photocuring; Fundamentals and Applications*, Hanser Publishers, Munich (1995).

Rechel, C. (Ed.), *UV/EB Curing Primer: Inks, Coatings and Adhesives*, RadTech International, North America, Northbrook, IL (1995).

Singh, A., and Silverman, J. (Eds.), *Radiation Processing of Polymers*, Carl Hanser Verlag, Munich (1992).

Fouassier, J.-P., and Rabek, J. F. (Eds.), *Radiation Curing in Polymer Science and Technology*, Vols. I–IV, Elsevier, London (1993).

Pappas, S. P. (Ed.), *Radiation Curing Science and Technology*, Plenum Press, New York (1992).

Oldring, P. K. T. (Ed.), *Chemistry and Technology of UV and EB Formulations for Coatings, Inks and Paints*, Vols. 1–5, SITA Technology Ltd., London (1991).

Clegg, D. W., and Collyer, A. A. (Eds.), *Irradiation Effects on Polymers*, Elsevier, London (1991).

Hoyle, C. E., and Kinstle, J. F. (Eds.), *Radiation Curing of Polymeric Materials*, ACS Symposium Series 417, American Chemical Society, Washington, DC (1990).

Farhataziz, M. A. J., *Radiation Chemistry*, VCH, Weinheim (1990).

Allen, N. S. (Ed.), *Photopolymerization and Photoimaging Science and Technology*, Elsevier, Essex (1989).

Reiser, A., *Photoreactive Polymers: The Science and Technology of Resists*, John Wiley & Sons, New York (1989).

McLaughlin, W. L., Chadwick, K. H., McDonald, J. C., and Miller, A., *Dosimetry for Radiation Processing*, Taylor & Francis Group, London (1989).

Bly, J. H., *Electron Beam Processing*, International Information Associates, Yardley, PA (1988).

Randell, D. R. (Ed.), *Radiation Curing of Polymers*, Royal Society of Chemistry, London (1987).

Humphries, S., Jr., *Principles of Charged Particle Acceleration*, John Wiley & Sons, New York (1986).

Guillet, J., *Polymer Photophysics and Photochemistry*, Cambridge University Press, Cambridge, UK (1985).

Bradley, R., *Radiation Technology Handbook*, Marcel Dekker, New York (1984).

Pappas, S. P. (Ed.), *UV Curing: Science and Technology*, Vol. 2, Technology Marketing Corp., Norwalk, CT (1984).

Schiller, S., et al., *Electron Beam Technology*, John Wiley & Sons, London (1983).

Raffey, C. G., *Photopolymerization of Surface Coatings*, Wiley-Interscience, New York (1982).

Pappas, S. P. (Ed.), *UV Curing: Science and Technology*, Vol. 1, Technology Marketing Corp., Norwalk, CT (1978).

Kase, K. R., and Nelson, W. R., *Concepts of Radiation Chemistry*, Pergamon Press, New York (1978).

Dole, M. (Ed.), *Radiation Chemistry of Macromolecules*, Academic Press, New York (1974).

Wilson, J. E., *Radiation Chemistry of Monomers, Polymers and Plastics*, Marcel Dekker, New York (1974).

Dole, M. (Ed.), *Radiation Chemistry of Macromolecules*, Vols. 1 and 2, Academic Press, New York (1972, 1973).

Holm, N. W., and Berry, R. J., *Manual on Radiation Dosimetry*, Marcel Dekker, New York, (1970).

Chapiro, A., *Radiation Chemistry of Polymeric Systems*, Wiley-Interscience, New York (1962).

Charlesby, A., *Atomic Radiation and Polymers*, Academic Press, London (1960).

Appendix II: Major Equipment Manufacturers

TABLE II.1

UV Equipment Manufacturers

Manufacturer	Web Page	Products
American Ultraviolet Company	www.americanultraviolet.com	UV equipment, UV light sources, spot curing equipment, ballasts
AMS Spectral UV	www.amsspectraluv.com	UV equipment, UV-LED light sources, measuring equipment
ClearTech Industries Inc.	www.cleartech.ca	UV equipment, UV light sources, measuring equipment
Dymax Corporation	www.dymax.com	UV equipment, UV light sources, measuring equipment
EIT Instrument Markets	www.eitinc.com	UV equipment, testing equipment
Excelitas Technologies Corp.	www.excelitas.com	UV curing equipment, UV-LED light sources
EXFO Life Sciences and Industrial Division	www.company7.com/exfo	UV equipment, UV light sources, measuring equipment
GEW (EC) Ltd.	www.gewuv.com	In-line curing UV equipment
Hamamatsu Corporation	www.ushamamatsu.com	UV equipment, light and radiation sources, sensors, optical components, EB equipment
Hanovia	www.hanovia-uv.com	UV equipment, UV light sources
Heraeus Noblelight	www.heraeus-noblelight.com	UV equipment, UV light sources (standard, excimer, LED), ballasts

(Continued)

TABLE II.1 (*Continued*)

UV Equipment Manufacturers

Manufacturer	Web Page	Products
Hologenix, Inc.	www.hologenix.com	UV equipment, UV light sources, spot and small area curing systems, LED illumination
Honle UV America, Inc.	www.honleuv.com	UV equipment, UV light sources, testing equipment
IST Metz GmbH	www.ist-uv.com	UV equipment, UV light sources, testing equipment
Miltec UV	www.miltec.com	In-line UV curing equipment
Nordson UV	www.nordsonuv.com	UV equipment, UV light sources, ballasts, power supplies
OnLine Energy, Inc.	www.onlineenergy.com	UV equipment, UV light sources
Phoseon Technology	www.phoseon.com	UV equipment, LED illumination
Prime UV-IR	www.primeuv.com	UV equipment, UV light sources, testing equipment
UV III Systems Inc.	uv3.com	UV equipment, standard, 3D
UltraViolet Connection	www.ultravioletsystems.com	UV equipment, UV light sources, measuring equipment
Ushio America, Inc.	www.ushiosemi.com	UV equipment, UV light sources, UV excimer lamps
UVEXS Inc.	www.uvexs.com	UV equipment, measuring and testing equipment
UV Robotics LLC	www.uvrobotics.com	UV equipment and light sources, robotic systems with UV lamps
XENON Corporation	www.xenon-corp.com	UV equipment, UV light sources

TABLE II.2

EB Equipment Manufacturers

Manufacturer	Web Page	Products
Comet AG	www.ebeamtechnologies.com	Electron beam equipment, several designs
Elektron Crosslinking AB	www.crosslinking.com	Electron beam equipment, dose measuring equipment
Energy Sciences, Inc.	www.ebeam.com	Low- to medium-voltage electron beam equipment
Hamamatsu Photonics	www.hamamatsu.com	Radiation equipment

(*Continued*)

TABLE II.2 (*Continued*)

EB Equipment Manufacturers

Manufacturer	Web Page	Products
IBA	www.iba-worldwide.com	Electron beam and X-ray irradiation equipment
METALL+PLASTIC	www.metall-plastic.com/en	Electron beam curing and sterilization equipment
Nissin Electric	www.nissin.co.jp	Electron beam and X-ray irradiation equipment, several designs
PCT Ebeam and Integration LLC	www.teampct.com	Low- to medium-voltage electron beam equipment
Steigerwald Strahltechnik GmbH	www.steigerwald-eb.de	EB equipment mainly for welding, drilling, and research

Appendix III: Standard Radiation and Dosimetry Tests

ASTM E170: Standard Terminology Relating to Radiation Measurements and Dosimetry

ASTM E668: Standard Practice for Application of Thermoluminescence-Dosimetry (TLD) Systems for Determining Absorbed Dose in Radiation-Hardness Testing of Electronic Devices

ASTM E2232: Guide for Selection and Use of Mathematical Methods for Calculating Absorbed Dose in Radiation Processing Applications

ASTM E2303: Standard Guide for Absorbed-Dose Mapping in Radiation Processing Facilities

ISO/ASTM 51026: Standard Practice for Using the Fricke Reference-Standard Dosimetry System

ISO/ASTM 51204: Standard Practice for Dosimetry in Gamma Irradiation Facilities for Food Processing

ISO/ASTM 51261: Standard Guide for Selection and Calibration of Dosimetry Systems for Radiation Processing

ISO/ASTM 51275: Standard Practice for Use of Radiochromic Film Dosimetry System

ISO/ASTM 51276: Standard Practice for Use of Polymethylmethacrylate Dosimetry System

ISO/ASTM 51310: Practice for Use of a Radiochromic Optical Waveguide Dosimetry System

ISO/ASTM 51400: Practice for Characterization and Performance of a High-Dose Radiation Dosimetry Calibration Laboratory

ISO/ASTM 51401: Practice for Use of a Dichromate Dosimetry System

ISO/ASTM 51538: Practice for Use of the Ethanol-Chlorobenzene Dosimetry System

ISO/ASTM 51607: Practice for Use of an Alanine-EPR Dosimetry System

ISO/ASTM 51608: Practice for Dosimetry in an X-Ray (Bremsstrahlung) Facility for Radiation Processing

ISO/ASTM 51631: Practice for Use of Calorimetric Dosimetry Systems for Electron Beam Dose Measurements and Routine Dosimeter Calibrations

ISO/ASTM 51649: Practice for Dosimetry in an Electron-Beam Facility for Radiation Processing at Energies between 300 keV and 25 MeV

ISO/ASTM 51650: Practice for Use of a Cellulose Triacetate Dosimetry System

ISO/ASTM 51701: Guide for Performance Characterization of Dosimeters and Dosimetry Systems for Use in Radiation Processing

ISO/ASTM 51702: Practice for Dosimetry in Gamma Irradiation Facilities for Radiation Processing

ISO/ASTM 51707: Guide for Estimating Uncertainties in Dosimetry for Radiation Processing

ISO/ASTM 51818: Practice for Dosimetry in an Electron Beam Facility for Radiation Processing at Energies between 80 and 300 keV

ISO/ASTM 51940: Guide for Dosimetry for Sterile Insect Release Programs

ISO/ASTM 52116: Practice for Dosimetry for a Self-Contained Dry-Storage Gamma-Ray Irradiator

Other Resources for Dosimetry

Guidance Notes for Dosimetric Aspects of Dose Setting Methods, Panel on Gamma and Electron Irradiation, London (1997).

Guide to UV Measurements: UV Technology, Radiometry and Measurements for Industrial UV Curing, RadTech International North America, Chevy Chase, MD, www.radtech.org. CD ROM.

Sharpe, P., and Miller, A., Guidelines for the Calibration of Dosimeters for Use in Radiation Processing, National Physical Laboratory, Teddington, UK (1999).

Raymont, J. and Kashyap, A., Measuring the Output of Ultraviolet Light Emitting Diodes, *RadTech Report*, Spring 2011, www.radtech.org.

LEDCure™ Brochure PN0104, August 2019, EIT®, www.eit.com.

Other Tests for Radiation-Cured Coatings

Guidelines for the Development, Validation and Routine Control of Industrial Radiation Processes, IAEA Radiation Technology Series No. 4, International Atomic Energy Agency, Vienna (2013).

ASTM Standards Related to Testing of Radiation-Cured Coatings, ASTM International, West Conshohocken, PA (2002).

Appendix IV: Major Suppliers of Raw Materials for UV/EB Curing

Supplier	Web Page
Allnex	www.allnex.com
Arkema	www.arkema.com
BASF	www.basf.us/dpsolutions
BCH	www.bch-bruehl.com
Covestro	www.covestro.com
Deco-Chem	www.decochem.com
DuPont Powder Coatings	www.dupontpowder.com
Dymax Corporation	www.dymax.com
Esstech Inc.	www.esstech.com
Excelitas Corporation	www.excelitas.com
Humpford Research	www.humpfordresearch.com
IGM Resins, B.V.	www.igmresins.com
Lambson Ltd.	www.lambson.com
Lubrizol Advanced Materials	www.lubrizol/coatings.com
RAHN AG	www.rahn-group.com
San Esters Corp.	www.sanesters.com
Spectra Group Limited, Inc.	www.sglinc.com
Sartomer	www.sartomer.com
Sun Chemical	www.sunchemical.com
Velsicol Chemical LLC	www.velsicol.com

Appendix V: The Twelve Principles of Green Chemistry*

1. It is better to prevent waste than to treat or clean up waste after it is formed.
2. Synthetic methods should be designed to maximize the incorporation of all materials used in the process into the final product.
3. Whenever practicable, synthetic methodologies should be designed to use and generate substances that possess little or no toxicity to human health and the environment.
4. Chemical products should be designed to preserve efficacy of function while reducing toxicity.
5. The use of auxiliary substances (e.g., solvents, separation agents, etc.) should be made unnecessary whenever possible and innocuous when used.
6. Energy requirements should be recognized for their environmental and economic impacts and should be minimized. Synthetic methods should be conducted at ambient temperature and pressure.
7. A raw material feedstock should be renewable rather than depleting whenever technically and economically practical.
8. Unnecessary derivatization (blocking group, protection, and deprotection; temporary modification of physical and chemical processes) should be avoided whenever possible.
9. Catalytic reagents (as selective as possible) are superior to stoichiometric reagents.
10. Chemical products should be designed so that at the end of their function they do not persist in the environment and break down into innocuous degradation products.
11. Analytical methodologies need to be further developed to allow for real-time, in-process monitoring and control prior to the formation of hazardous substances.
12. Substances and the form of a substance used in a chemical process should be chosen so as to minimize the potential for chemical accidents, including releases, explosions, and fires.

*Anastas, P. T., and Warner, J. C., *Green Chemistry Theory and Practice*, Oxford University Press, Oxford, UK, p. 30 (1998). (With permission from Oxford University Press.)

Index

Note: Page numbers in bold and italics refer to tables and figures, respectively.